'The automobile era is giving way to a new form of networked mobility, driven by digital technology but involving everything from new forms of transportation and electric, driverless vehicle to bicycles and our two feet. In this engaging and important book, Rossant and Baker tell the eye-opening story of this mobility revolution and what it means for our society, our planet, and each and everyone one of us.'

RICHARD FLORIDA, author of *The Rise of the Creative Class*

'Mobility is about us: enabling billions of people on the move with a new promise for freedom and choice. John Rossant and Stephen Baker tell us stories of 3D printed cars assembled by robots, accessible at your fingertips through ubiquitous apps and moving autonomously guided by networked devices. Mobility is the killer app for the fourth industrial revolution.'

PROFESSOR KLAUS SCHWAB, Founder and Executive Chairman of the World Economic Forum

'How we live in cities – how we experience cities – is a direct reflection of how we move around. In fact, a city's very identity is intimately connected to transportation – from the Paris Metro and London's double-deckers and black cabs, to the freeways of Los Angeles. *Hop, Skip, Go* paints an extraordinary portrait of the profound revolution underway in mobility – and the dramatic impacts this will have on the twenty-first-century city.'

DANIEL LIBESKIND, architect

HOP SKIP GO

ALSO BY STEPHEN BAKER

The Numerati

Final Jeopardy:
Man vs. Machine and the Quest to Know Everything

The Boost

HOP SKIP GO

HOW THE TRANSPORT REVOLUTION *IS* TRANSFORMING *OUR* LIVES

JOHN ROSSANT *AND*
STEPHEN BAKER

HarperCollins*Publishers*

HarperCollins*Publishers*
1 London Bridge Street
London SE1 9GF

www.harpercollins.co.uk

First published by HarperCollins*Publishers* 2019

1 3 5 7 9 10 8 6 4 2

A catalogue record of this book is available from the British Library

ISBN 978-0-00-830948-0

Printed and bound in Great Britain by CPI Group (UK) Ltd, Croydon

MIX
Paper from responsible sources
FSC™ C007454

This book is produced from independently certified FSC™ paper to ensure responsible forest management.

For more information visit: www.harpercollins.co.uk/green

To the young ones,

Alex Rossant, Sloane Wright, and Louise Craver,

who will have so many different choices

when it comes to moving around.

CONTENTS

INTRODUCTION

Where have you been your whole life? Think of all the places: home, school, work, restaurants, the gym. Maybe Carnival in Rio, a cheap hotel in Paris? All those memories and routines involved moving your body from one place to another. Movement is such a constant in our lives that it's easy to overlook. It occupies hours of each day and devours a good share of our money. It's central to our existence, nearly on a par with eating. If we didn't move, we'd never see each other. Now, as we enter the third decade of the twenty-first century, the ways we move are changing dramatically.

At about this point, as we describe this book, people often interrupt and say, "Driverless cars, right?"

What's coming is more than just one instance of robotic navigation, and much bigger. You can see it gearing up around the world. Overwhelmed by traffic and wheezing from smog, cities are confronting the limits of the century-old automobile economy and taking steps to rein it in. Many are extending greenways and building bike lanes. Some, like Madrid, are starting to ban cars altogether from city centers. Others are charging fees to drive in congested areas while presiding over a proliferation of bike- and car-shares. At the same time, a technology explosion is unleashing a new generation of networked machines and services. And yes, it's true, the automobile, step-by-step, is turning into a robot.

The car we all know, the gas-gulping machine we drive, has had quite a century. Our urban geography has molded itself to its demands. Its networks of highways gave birth to new suburbs, while developers carved out vast downtown acreages for parking lots. Cities grew into planned preserves for countless herds of these four-wheeled creatures, Serengetis for cars and trucks. That automotive monoculture spread around the world, from Los Angeles and Caracas to Moscow and, more recently, Beijing and New Delhi. The car reshaped the surface of the planet and poisoned the air above.

While the automobile isn't going away, it is losing its primacy. Over the next decade, many of us will find new ways to move. This next stage of human mobility, unlike the last, will not be defined by a single iconic technology. Instead we'll be faced with a host of choices, most of them tracked and mediated by digital networks. Where it works, this arrangement will be greener, cheaper, faster, and safer than today's smoggy and gridlocked status quo. The financial impact, for everything from households to national economies, will be profound. Once again, mobility will reshuffle our industrial landscape, juggle the timetables of our lives, and reshape our cities. It will also weigh heavily, for better or worse, on the future of our planet.

Much of the change ahead is tied to digital knowledge—a sprouting awareness of where we are, where we want to go, and what's available to take us there. This intelligence poses a profound contrast to the status quo of the automobile age, which has been marked by abject ignorance.

To recall just how little we knew, picture the outlaws Bonnie and Clyde on their murderous bank-robbing rampage through the Great Plains in the 1930s. They were driving on public highways, *yet no one knew where they were.* The police had to track down eyewitnesses and then stick time-colored pins in maps as they attempted to determine where the outlaws might be and where they'd strike next. Cars, for most of the century, were as footloose as hyenas. For millions of teenagers, this ignorance of whereabouts spelled freedom. We were out driving . . . somewhere. Our parents trusted we weren't robbing banks. But they knew only vaguely, if at all, where we might be.

City planners were also in the dark. Until recently, highway engineers would stretch cords equipped with counting sensors across roads. This provided primitive insights, but only on the behavior of our motorized herds. And engineers had precious few tools with which to manage the flow. Traffic, like a force of nature, simply happened. So most cities bowed to its unrelenting pressure and dedicated big chunks of their budgets to widening roads and highways, and opening up more space for parking. The resulting status quo was rife with waste—of energy, time, and money. The massive backups, the crawling lines of single drivers in five-thousand-pound hunks of metal, epitomized its inefficiencies.

The abundant waste found in the ignorant status quo spells boundless opportunity. Planners and entrepreneurs alike can now measure and count all kinds of things. As they do, they can sharpen their focus from the herd to the individual. This fuels new businesses. A commuter in Paris, for example, inches along the clogged Boulevard Périphérique at the wheel of a Renault Clio. Say there's a service or an app that can slice a half hour off her commute, in each direction, freeing up five hours a week. What would she pay for those miles and minutes?

These are burgeoning markets of space and time. Most of them would have been beyond our reach until recently. But a host of enabling technologies are now at hand. Arrays of sensors can now provide second-by-second reads on our location. High-speed networks, including 5G, can zip that data to computing clouds. Advanced artificial intelligence can turn it into insights about where we're going—and how best to get us there. Blockchain and other distributed ledger technologies, devised only a decade ago as the foundation for cryptocurrencies like Bitcoin, provide secure channels through which vehicles can share information and coordinate their actions. This sharing is fundamental for creating hyperefficient networks—even while we're still behind the wheel.

One other fundamental technology is the battery. The new mobility machines run, for the most part, on electricity, which happens to be what computers eat. Most of the new conveyances, in essence, are networked computers fitted with wheels or wings. Even bikes are getting smart.

Mobility, as we see it, represents the third stage of the Internet. The networked age debuted in the final decade of the last century, when rich troves of information, from mail to music, went digital. In the following decade, with the explosion of smartphones, information went mobile. That was stage two. The marriage of the cell phone to a networked computer placed the universe of knowledge into our pockets and purses, and added what amounts to an electronic lobe to the human brain. We're now accustomed, and even addicted, to carrying around networked technology. As we embark on stage three, ever more advanced blends of that technology are *carrying us around*.

In the coming age of mobility, practically every rolling and flying conveyance will be a networked device. For a glimpse of what this means, open a ride-sharing app, such as Uber or China's DiDi. The dots on the screens provide a glimpse of the future, one in which each of us moves as a node on a network map.

Ride-sharing was the first big splash of a tech wave that had been hyped for two decades: the so-called Internet of Things. Refrigerators would order milk, we were told, and smart light bulbs would brighten when we walked into a room. A good number of these machine-to-machine applications are finally creeping into the marketplace. But the biggest and richest arena for the Internet of Things will be mobility. It dwarfs the rest of the field.

This doesn't mean that the next cars we buy will be driverless, or that we'll soon be gliding over rush hour traffic in autonomous flying pods. These technologies, as we'll see, are in the works and will indeed be lifting off, in some places much sooner than others. For most of us, though, the coming transition will begin not with a miraculous new machine at the curb, but with a series of new questions.

Imagine you're a twenty-nine-year-old, just out of grad school, taking your first good job. You're in LA. The commute—the only real downside to the job—looks to be a fifty-minute rush hour crawl to and from Santa Monica, most of it on the I-10. You need a car.

The first thought is money. For an apartment renter like you, the car will be far and away your most valuable possession. You'll need a loan to buy it. Factoring in price and maintenance, it will cost about $8,000 a year to own and maintain the car. *And the dumb thing will be parked for about 95 percent of its existence.* It's a cash drain. Even parking it costs money. Car ownership isn't all that different from caring for a needy animal and paying for regular visits to an expensive vet. In fact, if you consider expense and upkeep, and even function, the car has much in common with the high-maintenance mobility heavyweight it dethroned, the horse.

It's at this point, with an auto dealer reaching for your wallet, that you might consider the shape of the coming decade you'll be sharing with this hypothetical car. Will you really need it?

Try thinking about it the other way: Instead of a car, you have the $8,000 a year that owning it would cost. That's your transportation fund. Even a decade ago, the money wouldn't have gone far. There was no Metro train between downtown LA and Santa Monica. Bus service was a bad joke. Taxis—if you could find one in LA—were out of the question; they'd eat up $8,000 in a few months. The only choice back then was whether to buy a car or to lease one.

But the next decade will offer new choices, and more of them every year. Ride-sharing services and scooters are just the beginning. Los Angeles Metro, once an afterthought for the hordes on the highways, is expanding dramatically, thanks to billions of dollars in bonds approved by voters. Drivers, it turns out, were so desperate for relief that they funded transit by raising their own gas tax. For many, it's a prayer that better public services will deliver a dose of relief for highway traffic and make driving tolerable once again.

The question remains: Do you spring for the car? While your options will continue to expand, complications remain. What do you do, for instance, when winter rains pelt down and the scooter ride from the Metro station is in the dark?

The answers still aren't easy. Even as new networks and technologies take root over the coming decade, many of us will stick with the car—or be stuck with one. The mistake, though, is to expect that the next stage of mobility will be like the last, simply with smarter and cleaner machines. In this vision, we keep the driveway, the two-car garage, and the parking lots. We hold on to the entire automobile monoculture, but we replace the gas-powered machines we've been driving for the last century with zippy electric models that will eventually drive themselves.

It's natural to think that way. Early in the twentieth century, many people assumed that automobiles would simply replace horses. Sure, they'd consume gas instead of oats, foul the air more than the roads, and run faster. And yes, when they broke down, it would be a mechanic, not a veterinarian, who'd coax them back to health. But the world would continue more or less the same. The name for these new machines said as much. They were "horseless carriages."

The automobile, however, went on to mold much of the global economy, from manufacturing to oil. It also transformed urban geography. Over the last century, asphalt, a byproduct of oil refining, coated a good chunk of the planet, including the parts where most of us live. Asphalt is what cities wear.

Billions of us are still tied to cars and trucks. Some of us adore them. Others drive simply because in the world created by cars, the car remains the best way to get from here to there. That's why automobiles have essentially run the world, and much of its economy, for the last century.

But now the entire car economy, from giants like Toyota and Ford to Daimler, has to find its place in a world of connected vehicles, most of them running on electricity. Oil companies also face existential challenges. The move away from the internal combustion engine threatens to eviscerate the market for gasoline, which currently accounts for nearly half of the oil we consume. That could lead to a collapse of oil prices. The need for lithium and a host of rare earth minerals, all of which are key battery components, could spark mad rushes to Chile and the Democratic Republic of Congo. Disruption awaits everywhere.

Consumers, meanwhile, will be facing a smorgasbord of new possibilities. A host of start-ups, such as companies like Arcimoto, in Oregon, are building new types of mobility machines. Arcimoto's electric buggy is a three-wheeled blend of a car and cycle, with a roof. In China's provincial capitals, the streets teem with bike-like microcars. They're slow, with a range of only seventy miles, and they have as much protective armor as a toaster. But some of them cost only $1,000. Mobility at its most basic level is going to get extremely cheap.

This shift toward networked mobility will also stir up social questions, many of them fraught. While the automobile economy created and still sustains tens of millions of middle-class jobs, AI and robotics enrich a much smaller technology elite. The migration of mobility toward the digital economy threatens to widen the global gap between haves and have-nots. What's more, cities are the laboratories for this revolution. If new mobility options enhance their competitiveness and lift the urban quality of life, they're likely to attract even more talent—in turn exacerbating the city-country divide.

THIS STORY OF movement is almost as big as the earth itself. So how do we tell it?

We start with the people who are putting it together—creating the flying machines, electric cars, AI-fueled services, and new networks that will change the way we move. The people we'll meet in each chapter represent entire ecosystems of activity. There's Kevin Czinger, the founder of Divergent 3D. Headquartered just a few miles south of LA's airport, Divergent is building an entirely new automobile manufacturing setup. The car is designed on a computer, its parts spit out by 3-D printers and then assembled largely by a team of robots. If you're not happy with it, you melt the pieces, tweak the design, and try again. That's the advantage of a software economy. It's easy to change and to try things. It's why innovation comes so fast. The world of mobility is facing whirlwinds of it.

We'll also explore one of the most ambitious education efforts currently under way in artificial intelligence: the race on three continents to teach machines how to drive. We'll see a

Chinese cartographer named Wei Luo, in a Palo Alto start-up, DeepMap. She's building a new generation of maps, which direct autonomous cars by the centimeter and alert them to fender benders and fallen branches. This is a new rendering of geography for a different and infinitely more exacting consumer: a machine. Meanwhile, in China's southern megapolis of Guangzhou (once known as Canton), scientists at a start-up called Pony.ai are fine-tuning the cognitive processes of autonomous cars running down their streets, while trying to figure out how much of what they learn in China, or in California, will work in Rome or Cape Town.

Building systems that can navigate the physical complexity of the world by themselves, while also dealing with swarms of impetuous, irrational, jaywalking, drive-texting humans, will be one of the towering technological feats of the century. For most of the coming decade, though, the lion's share of business in autonomy will be for applications that can make money while smart cars are still undergoing their apprenticeship. In other words, while full autonomy is the research goal, *semiautonomy* will be the bigger market, at least over the next few years.

Robert "RJ" Scaringe, founder of the electric car start-up Rivian, whom we'll meet in chapter 3, imagines a semiautonomous feature for his coming electric SUVs and pickups: a guided autonomous tour of a national park—maybe Yellowstone or South Africa's Kruger. The car will drive itself on a premapped course, as if on train tracks, and provide commentary en route. This will not be full autonomy, but it should offer enough of it for the driver to turn around and gape at a bear or a giraffe, and even to take pictures while the wheel turns itself. This apprenticeship will be valuable not only for the cars (or whatever we end up calling them). It will also introduce us to autonomy little by little, so that when machines are finally fully capable of navigating the surface of the earth, we might not find them scary. Maybe by then we'll have figured out how to regulate them.

Not all the new movement will be happening on land. A host of companies, from start-ups to aviation titans like Airbus and Boeing, are engineering a new generation of flying machines. An engineer named Mark Moore worked for three decades at NASA to develop robotic flying craft. For much of that time, they seemed less likely than the robots destined for Mars. Yet now more than one hundred companies around the world are building these electric airships and whirlybirds, and Moore has jumped from NASA to Uber, where the plan is to operate networks of robotic flying taxis in cities, starting with Los Angeles and Dallas. Flying pods sound even more exotic than driverless cars, and yes, there are all sorts of legal and regulatory wrinkles to iron out. But once these vessels are up in the air, they face a simpler mission than that of autonomous cars, simply because the sky is a lot less busy than our tangled world below.

Other mobility ventures are driven less by cutting-edge technology than by imagination and entrepreneurship. In 2010, Nadiem Makarim, a young Indonesian, had the idea to organize so-called ojeks, or motorcycle taxis, into a ride-sharing network, Go-Jek. He launched in Jakarta with twenty of them, and now Go-Jek has more than a million of these two-wheeled chauffeurs, and the company is spreading throughout south Asia. Once Makarim had his mobility app on millions of phones, he started to use it to deliver a host of other services, from banking and massages to package delivery. His company runs on data, and it operates one of the biggest AI labs in Southeast Asia. Little wonder Google recently led a $1 billion investment round for a stake in it.

THIS THIRD STAGE of the digital age differs from its PC and smartphone predecessors in two fundamental aspects. The first has to do with geography. The first network boom was largely local: a handful of companies on the West Coast of the United States defined the Internet revolution and towered above it. The dominant smartphone platforms—those produced by Apple and Google—came from the same neighborhood. But the mobility revolution, from its very inception, has been a global phenomenon. China is on equal footing with the United States, if not ahead of it. Israel is a key technology player, as are European hot spots like Berlin. Entrepreneurs in Nigeria, Indonesia, and Mexico are creating transformative networks that can catapult them into international markets. In short, the mobility revolution is exploding everywhere.

With one caveat: it starts in cities. And that leads to the second big difference in mobility. While the Internet engendered its own virtual worlds on screens, mobility takes place in the physical realm we inhabit, much of it in our shared space. It involves roads and bike lanes, and machines that can hurtle down sidewalks and bang into us. It has to be managed. So from the get-go, there's a clear role for government.

It stands to reason that the world's cities would serve as laboratories for new forms of mobility. Only they can offer thousands of potential customers within a few blocks of one another. That's why it's a cinch to summon a ride-share or rent a bike in Munich or Tokyo, and why it's impossible in most of Patagonia or Wyoming. Cities are also central to our story because, increasingly, humanity is turning into an urban species. Today, more than half of us live in cities. That number is expected to grow to 68 percent of the global population, topping 9.5 billion, by 2050. This means that the urban population is expanding at an average of 1 million people—or a city the size of Stockholm—every week. Facing this crush, many cities will find themselves looking to smart mobility as an antidote to urban paralysis and asphyxiation.

But still, people will have to figure out how to manage it. That's why, in the course of telling this story, we take to the road (and the skies) to visit four cities on four continents. Each one is a petri dish for new ways of moving people and their things. Each will come up with its own formulas (while borrowing liberally from hundreds of other cities around the world). Their decisions, from taxes to investments in electric chargers, new train lines, and the rules of the road for autonomous fleets, will shape neighborhoods, define commutes, and establish competitiveness worldwide. After all, if a company has a choice between two cities, one gridlocked and smoggy, the other where movement is faster, cheaper, and greener, the decision is a no-brainer.

You might think that Los Angeles, the most iconic car city on the planet, would be the last place for a mobility revolution to occur. For starters, who can organize it? The mayor not only shares power with the city council, but he or she must also grapple with no less than eighty-seven other independent and headstrong municipalities within the vast LA County, with a population larger than forty-one individual American states. Yet as we'll see, LA is positioning itself as a sprawling test bed for what's coming. It has powerful assets: a sunny climate that smiles on scooters and bikes, an explosion of mobility tech start-ups, and also a leadership position in the aerospace industry. (A mastery of new materials and lightweight construction is highly relevant to the new contraptions we'll be riding.) Perhaps most important, LA has a population that's fed up with jammed freeways and ready for new approaches.

Later, we'll visit Dubai, where a government flexed with power and a near-limitless budget is positioning itself as a mobility pioneer. The plan is to operate flying taxi drones by the early 2020s. By 2030, according to the government's plans, one-quarter of the kilometers traveled in Dubai will be autonomous. And a planned vacuum train—a so-called Hyperloop—promises to shrink the two-hour drive to neighboring Abu Dhabi to a mere fourteen minutes.

Compared with LA and Dubai, Helsinki already seems like a mobility dream. It's a walkable city with trams, subways, and buses that seem to pass by every minute or two. But it still wasn't easy, a decade ago, for a college student named Sonja Heikkilä to cross town for soccer practice. Her frustrations grew into a vision: What if she could call up every available form of transportation on a single smartphone app—and then pay for mobility as a subscription, as with Spotify or Netflix? Her idea is taking root in Helsinki, where owning a car could one day become an anachronism, as quaint as buying a CD. (Auto executives need look no further than the beleaguered music industry to see how disruptive the transition from physical ownership to a digital service economy can be.)

Perhaps the most important stop on our global mobility tour is Shanghai. Until a few decades ago, it was a city of buses, bicycles, and swarms of pedestrians. Cars were rare. But now, like the rest of China's industrial east, Shanghai is transformed. It's become a thriving megapolis of twenty-five million, complete with jammed freeways and dirty air. The city's soaring ambition is to solve these problems while leading the world in the race to new forms of mobility. Its greatest asset is a bottomless ocean of data describing the movements and moods of every person in the city. This data, combined with advanced AI in the hands of a powerful government and a single dominating Communist Party, could one day manage human movement with the precision of an industrial supply chain.

But is this the future we want, each of us being shuttled from place to place with remorseless efficiency, like the fenders, seat cushions, and spark plugs in a Toyota plant? With each advance of the mobility revolution, we'll confront tough choices about the kind of world we'd like to live in. On one side will be the freedom of the individual. This includes the freedom to be wasteful, to travel incognito, even to get lost. Balanced against that are the interests of society, such as safety, efficiency, economic competitiveness, and a clean environment. This tension between the individual and society will run through the mobility revolution, and through this book. The contrast will come into stark relief as we explore the race to mobility in the world's two leading economies, those of the United States and China.

In the end, the story is about us: billions of people on the move. We measure each trip we take, whether to Kuala Lumpur or the corner drugstore, in the distance we travel, the time it takes to get there, and the money it costs. Miles, minutes, dollars. Underlying each stage of mobility, from foot to horse to cars and jets, are the mathematics of these three fundamental variables: time, space, and money. The coming mobility revolution will be no different. In the conclusion, we'll explore how the coming changes might revamp our conception of the geography of our planet, the hours in our days, and where in the world we might be able to go.

HOP SKIP GO

1

Hit Enter to Print Car

If you wanted to pinpoint the epicenter of movement in the United States, you might consider the otherwise unremarkable Los Angeles suburb of Torrance, California. It's barely ten minutes south of Los Angeles International Airport, the fourth-busiest in the world. The ports of Los Angeles and Long Beach, which together handle more than one-third of the cargo coming into the United States, are just a couple of exits away. In the part of Torrance where Kevin Czinger has set up shop, near the junction of two teeming ten-lane interstates, the 110 and 405, the wide boulevards seem endless. They're flanked by warehouses and dominated by trucks. The skinny palm trees in the medians look lonely and forlorn.

Czinger arrived at this hub in 2014 with a bigger-than-life goal: to reinvent auto manufacturing for the next century. The reigning status quo, as Czinger sees it, features massive factories mass-producing cars and trucks. These plants are geared toward tonnage and are punishingly rigid. The vehicles rolling off their lines choke our cities and cook our planet. It's a model, as Czinger sees it, well along the path to extinction. His alternative is designed to diverge from that death march, which is why Czinger named his company Divergent 3D. His model is fast and flexible, similar, in Darwinian terms, to the small, cagey mammals that survived cataclysmic climate change millions of years ago. Automakers large and small, in his view, need the same mammalian skills—speed and flexibility—to survive. That's what Kevin Czinger wants to sell.

Czinger's manufacturing setup, like so much else in the coming age of mobility, encodes the entire process in software. Like a magazine or a song, or a million other products in today's world, the entire vehicle is pieced together virtually, on a computer. Once the engineers are happy with the design, they hit the Print button, and 3-D printers spit out panels and joints, each one optimized for weight, strength, durability, fuel efficiency—in short, whatever quality the engineers ask for. Later a small cohort of robots assembles the car, plunks in an engine (either gas or electric), and adds four wheels and a few finishing touches. If the car is not quite right, the Divergent team melts down the pieces, tweaks the design, and reprints. As Czinger sees it, this new process will allow entrepreneurs and small design studios to barrel into small-scale car manufacturing. Setting up mini–manufacturing plants, he says, will cost a tenth of the mass-manufacturing norm, perhaps only $50 million, and will give birth to all sorts of boutique automakers. "We could have ten new car manufacturers in LA alone," he says. "[The outdoor retailer] Patagonia could make their own brand of cars."

It looks, though, like the first market for these 3-D printed cars will be in China. Czinger's leading investors, including the Hong Kong real estate magnate Li Ka-shing, are members of a Chinese syndicate. They've earmarked more than $100 million for his operation, and are setting up the first 3-D car manufacturing plant in Shanghai.

ON A SUNNY spring day in Torrance, Kevin Czinger strolls through his spacious office, past a few rows of programmers and engineers hunched over computers. He opens the big metal door onto a dusty construction site. A crew is busy leveling the ground for a team of car-building robots. The surface must be perfectly flat so that the robots can piece together the cars with millimetric precision.

Czinger, in his late fifties, has the erect bearing of a soldier. He wears a tight short-sleeved shirt. The arms coming out of the sleeves seem a couple of sizes bigger than the rest of his body, and both are popping with veins. He was a college football player, and in the terms of that sport, he looks like a defensive back with the arms of a lineman.

He grew up the youngest of five in a working-class family in Cleveland. Two of his brothers were mechanics and into drag racing. Czinger, while still in high school, refitted a '68 Plymouth Barracuda with a powerful 440 wedge V8 engine. That was hot-rodding. In those days, he says, it was the closest thing to computer hacking. "You had stuff from the manufacturers that didn't work so well, and you were trying to make something better."

It turned out to be football, not hot-rodding, that carried Kevin Czinger from Cleveland to Yale—and kick-started his career. Czinger was a demon on defense. Less than two hundred pounds, he played noseguard, lining up right across from the much-burlier centers. From this spot, just a few feet from the opposing quarterbacks, he waged war on the offense. "For three years," wrote the *Harvard Crimson* in 1980, "the key to the [Yale] defense's success has been middle guard Kevin Czinger." The article quoted the coach at Brown University, who said that Czinger had "singlehandedly" dominated his team two years in a row. "He ruins your whole game plan."

After college, Czinger signed up for the Marine Corps and went to law school, also at Yale. He later clerked for Gerhard Gesell, a federal judge who presided over momentous cases, from the Pentagon Papers to Watergate. When Czinger expressed an interest in prosecuting crime, Judge Gesell lined him up at the US District Court for the Southern District of New York, where Czinger soon found himself working as an assistant US attorney under Rudolph Giuliani. There he worked with a host of rising stars, including James Comey, the future FBI director.

Czinger's career reads like a guidance counselor's greatest hits—except that it's all one person. He won a prestigious Bosch fellowship in Germany, and he ran media and telecom for Goldman Sachs in London. For a couple of years, he was a top executive at the German publishing conglomerate Bertelsmann. When the Internet started to percolate as a history-changing force in the mid-1990s, Czinger landed in Silicon Valley, where, as chief financial officer and head of operations, he ran one of the most ambitious and visionary of the first wave of Web companies—and one of its most notorious failures. The first big online grocer, it was called Webvan.

Although Kevin Czinger's career continued its hopscotching path into this century, let's stop for a moment at Webvan, because it offers some parallels to his current venture, Divergent 3D.

There was a brief time, in the late 1990s, when Webvan had the look of a rising giant. Right after the 1995 initial public stock offering of Netscape, which set off the first Internet boom, a host of start-ups raced into what was then called "cyberspace" and carried out (online) land grabs. The thinking at the time was that each niche would eventually support only a couple of competitors, and perhaps only a single dominant player. Amazon, in these early days, was shaping up merely as the leading online bookstore. Webvan's aim was much grander. It would disrupt the $430 billion market for groceries. It promised to feed the nation, and with time, much of the world.

Venture capitalists showered millions on Internet start-ups, pretty much indiscriminately. Few of these dot-coms swallowed up bigger bets than Webvan. Leading venture firms, including the Silicon Valley titans Benchmark and Sequoia Capital, plowed hundreds of millions into the company. Webvan, intent on quickly establishing itself as the rising giant of e-grocers, rushed to open operations in ten major cities, including Chicago, San Francisco, and Los Angeles. This meant building high-tech warehouses and torturously complex supply chains, and delivering food to customers within a tight deadline—ideally, before it went rotten. It was insanely ambitious, and Kevin Czinger was calling most of the shots.

Czinger's current automaking venture is in many ways similar. It's ambitious and disruptive. It benefits from investors' rush into mobility. With Divergent 3D, Czinger continues to make bold promises. This time, instead of feeding the world, he's out to change the way we make things. Manufacturing, after all, is the foundation of the global economy. Snipping out 90 percent of the industrial process, as Czinger plans, could eliminate tens of millions of jobs. But it might also be the key to cleaning up humanity's ways—and carving out a path to sustainability.

Soaring visions like Czinger's flourish during the hype stage of a technological revolution, a period of sky-high promises and fearless investors. Visions sell and money flows. For many companies, profits are far off, and during this glass-half-full stage, that's OK. Each company can sell itself as a survivor, even a potential champion.

But Czinger's Webvan experience might tamp down a bit of the giddiness. There's a point in every boom, he knows, when markets turn from wonder to skepticism. This usually occurs after an early highflier goes bust, which leads investors to start asking hard questions about revenue and profits.

As pretenders fail, investors retreat and survivors feast on the fallen, picking up their code, their brainpower, their customers. In the aftermath, giants emerge. We cannot say at this point—no one can—where the companies we're getting to know, including Divergent 3D, will end up in this food chain. But whether they emerge as champions or fall and get swallowed along the way, they're busy building the next generation of mobility. Their workers and the code they write, and the industrial process they create, will play a part of this revolution, no matter which competitors emerge on top.

This was true for Webvan as well. The company was investing mightily, its monthly expenses dwarfing revenue. Profits, if they ever arrived, were years away. By the time the market's mood changed, in the spring of the year 2000, and investors' gazes turned from soaring visions to the bottom line, Webvan might as well have worn a sign around its neck saying SHOOT ME. The company's funding dried up, and it declared bankruptcy in 2002. Amazon, the survivor, promptly pounced on Webvan's assets, including its warehouses. That giant got bigger. "The core team up at Amazon," Czinger says, "is the old Webvan team, from robotics to warehousing." He and his team lost control, but fed the ecosystem.

Nonetheless, Kevin Czinger managed not only to escape this Darwinian drama, but to walk away a rich man. He had managed to vest rich options while Webvan's stock soared. He subsequently added to his wealth with lucrative stints in private equity.

So in 2008, not yet fifty years old, Kevin Czinger decided to do something big. His goal was nothing less than to save the world from global warming, while at the same time returning to his lifelong passion for cars. So he cofounded an electric car company, Coda Automotive. Most of his investors came from China, and the venture targeted the Chinese market.

As it turned out, Coda released its only model, an electric sedan, in 2012—the same year that Tesla came out with its hit luxury car, the Model S. The Tesla bested Coda in crucial categories, including range, and trounced it in the marketplace. That spelled Coda's doom. Within a year, Czinger was shoved out, and Coda was seeking bankruptcy protection. Now having succumbed to two iconic companies, Amazon and Tesla, Kevin Czinger was plotting his next move.

Even before Coda's fall, he says, he realized that his push for affordable electric cars was foolish—or at least dangerously misinformed. He had believed, like millions of electric car drivers today, that shifting the auto industry to an electric fleet would help save humanity from overheating our planet and killing ourselves. The cars don't pollute. They don't even have *tailpipes*.

But in 2009, Czinger came across a document that changed his thinking: a five-hundred-page report, *Hidden Costs of Energy*, produced by the National Academy of Sciences. It introduced the concept of life-cycle analysis and convinced Czinger that his entire vision (and those of other electric car companies, including Tesla) had everything backward.

Cars begin polluting, the report argued, long before a new owner presses the accelerator for the first time. It detailed the immense energy consumed in manufacturing a vehicle. This consumption starts with iron miners digging deep into the earth, hauling up mountains of ore, and loading them onto trains or barges. They transport it usually hundreds of miles, which consumes more energy. In steel mills, iron pellets fired with hard coal, called coke, are smelted in roaring blast furnaces that reach nearly 3,000 degrees Fahrenheit. The molten iron ore flows into other furnaces, where it's refined into thick slabs of steel, which are pressed by massive rolling pins and eventually flattened into sheets. Then the gleaming rolls of steel are shipped off to an auto plant. Each step of this process burns lots of fuel, in turn spewing metric tons of greenhouse gases into the atmosphere.

The other materials arriving at the same auto plant's docks—the plastics, glass, and chemicals—each emerge from their own industrial processes, most of them involving fires and furnaces. The manufacturing of a car, according to the report, consumes more energy and creates more earth-warming havoc than the actual car will produce as it plies the streets and highways for a decade or two. As Czinger read the report, it became clear to him that practically any new car, even an electricity-fueled Leaf or Tesla, was an environmental liability. "I was such a dummy," he says.

But this revelation led Kevin Czinger toward yet another staggeringly ambitious goal: This time, instead of feeding the world or electrifying transportation, he would take it upon himself to dramatically clean up auto manufacturing. And as if that weren't enough, he also aimed to minimize the environmental damage that cars create once they're built. This is where, a decade earlier, he had held out hope for emission-free electric cars. But the simple laws of physics overturned this logic. If you've ever tried pushing a car, you know that budging even a smallish one, a Mini Cooper, say, or a Camry, requires loads of energy (and a strong back). Compared with those cars, Tesla's luxury Model S, the one that sank Coda, was a behemoth. The first prototype, displayed in 2009, weighed in at 4,600 pounds. About one-quarter of that weight came from the battery alone. Moving millions of them would require countless gigawatts.

Even without an internal combustion engine, and absent the fumes, those gigawatts had to come from *somewhere*. About two-thirds of the electricity in the world, Czinger saw, came from burning fossil fuels. This added to global warming. Sure, there were promising trends. Norway's electrical grid was fueled by alternative energy. California was quickly making strides in that direction. Electricity in France came mostly from nuclear plants, which despite other concerns produce no greenhouse gases. But most of the world's electricity came from carbon, and it would for decades. Using that electricity to move heavy vehicles hundreds of billions of miles, Czinger realized, was unsustainable.

What's more, the world's biggest and fastest-growing car market—China—promised unmitigated disaster. Most of China's electric cars, in effect, would be running on filthy *coal*. This promised some relief for the coughing and wheezing masses in smoggy Beijing and Shanghai. But from a global perspective, it simply shifted the pollution from crowded cities to the distant fossil fuel–burning utilities elsewhere. For the future of the planet, it was even worse. "Turning China's fleet to electric cars," he says, shaking his head, "is the most insane thing you could ever think of doing."

So Kevin Czinger would not only clean up the industrial process. His manufacturing system would also produce dramatically lighter cars. At one-third the weight of traditional cars, they would consume less energy, regardless of the engine type. That would result in cleaner air.

His goal was to replicate the cyclical patterns and feedback loops of nature. This is a recurring theme throughout the mobility world, and indeed, in the broader sphere of computing. The idea is that throughout our industrial history, we have been starved of vital information, or feedback. Most traffic lights at four thirty a.m. don't see that we're waiting at the corner all alone. They cannot adjust to changing conditions, as a crossing guard might, and wave us through. Lacking this data, they blindly rely on programmed rules. On an intelligence scale, they're somewhere between rocks and refrigerators—unresponsive, but reliable. Entire industries, as we'll see, are focused on seeding elements of the physical world, including traffic lights, with sensors, and turning the dumb machines into adaptive networks, ones that behave more like plants and animals.

This same logic extends to manufacturing. Car companies spend hundreds of billions of dollars to mass-produce legions of identical units. It's a dumb, inflexible process. Lacking feedback loops to catch defects or to gauge popularity, an auto plant simply pumps out the units. If something's wrong, the company issues hideously expensive recalls. And if certain aspects of the car or truck—the hood design, lumbar support, highway mileage—turn off buyers, there's no easy fix. The structure is locked in. Failures cost hundreds of millions. It's money down the drain.

In Czinger's scheme, which he relates to biology, each car evolves. The 3-D printer process can spit out single specimens, which can be tested for speed, handling, comfort, fuel efficiency. This creates feedback loops. As test data comes in, the engineers can melt down the car and tweak the software design—the car's DNA. They can spawn different species for varying markets—or ecosystems—perhaps one car for the long, flat boulevards of Torrance, another for the chaotic streets and alleys of Karachi.

The former noseguard gets most excited when talking about the most violent of feedback loops: crashes. Because most of a printed car can be melted down and recycled, it's much cheaper to run them through crash tests. Each test will produce rich data on every material and design feature in the car. In Czinger's vision, next-generation manufacturers around the world will crash their cars, scores of them, creating oceans of feedback data, which they'll share with everyone else. "We'll be *swimming* in crash data," Czinger says. Once this data is fed into learning engines, they can analyze the performance of each component, gradually leading to the safest and most crashworthy designs.

Such is the supple nature and competitive advantage of a manufacturing process that exists, in large part, as software.

DIVERGENT 3D REPRESENTS merely one stab at manufacturing the next generation of vehicles. Entrepreneurs around the world are busy devising new machines, and a good number of them are innovating with schemes simpler than Kevin Czinger's robots and 3-D printers.

Many such start-ups are repurposing industrial machinery used to make stoves or bicycles. The result is an explosion of tinkering. VeloMetro, a Vancouver, British Columbia, start-up, created the Veemo, a three-wheeled electric-aided bicycle encased on three sides in an all-weather pod. Big companies are in on it, too. Renault's Twizy, a featherweight electric automobile, looks like a go-cart. Its two doors rise up on its sides, like a bat's wings. Over the coming decades, the streets and sidewalks of cities around the world will be crowded laboratories for a wild and diverse generation of mobility machines. They'll look like something dreamed up for video games, or from a world inspired by Dr. Seuss.

In the college town of Eugene, Oregon, a former video-game designer named Mark Frohnmayer is putting together one such machine, an electric auto–motorcycle hybrid called Arcimoto. Imagine, for starts, turning a tricycle around, so that two wheels are in front, one behind. Then expand it to the size of a motorcycle, put in a couple of seats, one behind the other, and enclose it with an arc of plexiglass. These odd beasts are now rolling out of an Oregon manufacturing plant and selling for $11,000.

Frohnmayer, a UC Berkeley–educated computer scientist, succeeded early in his career as a video-game designer. One of his hits in the late 1990s, *Starsiege: Tribes,* was an early online multiplayer game set a couple thousand years into the future. Each player's character was equipped with a gun, and he gathered with other tribes of humans for fights that jumped from one galaxy to the next. In 2001, Frohnmayer and his partners founded a software company called GarageGames. The idea was to develop easy-to-use tools for people to create their own video games. Six years later, Frohnmayer and his team sold the company to Barry Diller's Internet conglomerate, IAC, for a reported $80 million.

This left him with a chunk of money and some free time. So he went shopping for a car. After a successful "exit," as it's called in the venture business, plenty of entrepreneurs might splurge on a Tesla S or a Porsche Panamera. But Frohnmayer is in Oregon, not the Valley. He's the son of a university president, very idealistic, and, like many in the new mobility businesses, bright green and eager to save the world.

He was in the market for a socially responsible set of wheels, something to use when it was too wet to bike. He didn't want to spend too much for it—maybe $10,000. He was disappointed. Even the cheapest cars seemed too big and cumbersome. He considered a motorcycle. It would be easy to park and fuel efficient. But motorcycling is miserable in the rain, which pretty much defines Eugene from October to June. Also, motorcycles are dangerous. People fly off them like missiles.

"What I saw," Frohnmayer recalls, "is this enormous space between the motorcycle and the car." Most trips around town, he said, involve one person, sometimes two, rarely more. So there had to be a market for people like him, who wanted a cheap and extremely fuel efficient electric vehicle for driving around town in bad weather, maybe just the mile or two to a train station or bus stop. The vehicle, he thought, should be as easy to park as a motorcycle, but as safe as a car, and with enough storage to bring home a few bags of groceries. He figured he could assemble a team to design this new species.

It turned out to be harder than software. "When you build a game in software," he says, "you can copy it for no cost. You can fix a bug. Software is almost magical." Manufacturing in the physical world, by contrast, proved to be "fantastically more complex." Starting in 2008, his team in Eugene created one version after another of the two-seated roadster. This development went on for eight years—the entire Obama presidency. The Arcimoto team kept subbing in different materials and designs to reduce weight; they replaced handlebars with a steering wheel, then returned to handlebars again. They wedged stronger batteries into smaller nooks.

In all, they went through seven versions, and something was always . . . not quite right. But the eighth version sold them. It had a range of seventy-five miles. Though hardly a speed demon, the Arcimoto could still hold its own with cars, with a top speed of eighty miles per hour.

Finally, Frohnmayer had a green machine to strut before investors. In 2017, Arcimoto listed its stock on the Nasdaq Global Market and raised $19.5 million. That was enough to go into production. It launched sales early in 2019, but with a higher price than anticipated—some $19,900. Frohnmayer vowed to the press that with greater volume and expertise, they'd eventually get the price down to the $11,900 target. This is the tough learning curve start-ups face in manufacturing. The traditional players are wizards when it comes to mass production. No other industry has come close.

Arcimoto's manufacturing is primitive, compared with Divergent 3D's robotics and 3-D printing. In the Eugene factory, Arcimoto's workers shear metal parts from sheets of steel, and then use a press to bend them to the right shapes—"Essentially sheet metal origami," says Frohnmayer. The Arcimoto SRK, more motorcycle than car, is a far simpler vehicle than the cars Divergent 3D is designed to build. But Arcimoto also spent a lot less than the $50 million for a Divergent minifactory. The company has raised a mere $30 million in funding, and it already has its full manufacturing operation up and running. "And a lot of the money is still in the bank," Frohnmayer says.

In the fight to build the next generation of mobility manufacturing, a host of variables are in flux. The dollar investment in manufacturing is falling, as is the cost of vehicles. Meanwhile, the choice of vehicles is exploding, and each year batteries provide a greater range and lower costs. The challenge, whether the current product is an Arcimoto, a Veemo, a Twizy, or one of the 3-D printed cars coming off a new line in Shanghai, is to build an enduring business plan for times of unrelenting change.

AT THE DAWN of the Internet age, when Kevin Czinger was busy building an online grocery business, the mobility revolution would have been impossible. Yet over the following two decades, crucial technologies advanced dramatically, turning visions like 3-D printed cars from far-fetched fantasies into factory installations.

It's astounding how many crucial pieces have fallen into place in so little time. Start with data, the feedstock of the information economy. At the turn of the century, the age of data had not yet taken shape. This is because most of us weren't yet spending our lives interacting with screens and surrounded by sensors, or busily feeding social networks. The networks didn't learn much about us—our buying habits, our diseases, our networks of friends. Our lives were still largely off-line. Many computers, strange as it sounds today, sat cloistered in "computer rooms." Laptops had no Wi-Fi. And even if we had uploaded the data on floppy disks to networks, there weren't yet powerful cloud computers to store and process it, turning our behaviors and movements into insights and fueling crucial advances in artificial intelligence.

In those early Internet years, networked sensors, the eyes and ears of the mobility world, were still in their infancy. By far the most important of these sensors, the smartphones we carry around everywhere, did not yet exist. Without smartphones, an entire wing of the mobility economy, from Uber to dockless bike and scooter companies, would disintegrate. (For those companies, our smartphones are their customers, not us. We're simply stuff that rides along, the smartphones' luggage.)

One of the recent breakthroughs in the data economy has been the increased mastery of human language by machines. All our online scribbling and yammering has created massive language sets for computers. In effect, we have taught them language. This enables us to talk to the machines moving us around. Speech is the dominant interface for mobility technology. This wouldn't have been possible, except in a handful of primitive applications, before about 2015.

By looking back even a decade or two, we can sense the speed of the tech current pushing us forward. It's fast, and it's accelerating. The technologies powering the mobility revolution, from AI to manufacturing and network management, are sure to advance just as dramatically over the next decade or two.

The same growth curve is being experienced by 3-D printing. In the first decade of this century, the mere suggestion of harnessing armies of 3-D printers for automobile manufacturing would have sounded outlandish. Such printers in their infancy were mostly for hobbyists. Designers could draw up something on their computers—perhaps a refrigerator handle, or a new stem for a broken pair of sunglasses. But the process was slow. Similar to a child building a sand drip castle at the beach, the printer deposited material layer upon layer, and gradually an object rose into the physical world. It was called "additive" technology. Instead of a child's stubby fingers dripping sand, a 3-D printer used a precision nozzle that spit out minutely calibrated material, usually plastic. It was miraculous, in its way, but deliberate, built to craft one object at a time. It might have been the next stage of craftsmanship, but it was hardly a rival to mass manufacturing.

Yet the manufacturers of such devices kept shooting ahead. In the last decade, 3-D printers have radically expanded their diet and equipped themselves for more serious work. These digital factories can now consume a variety of metallic powders and a broad array of composites. This increases their range. Their speed, meanwhile, has climbed up the exponential curve familiar to other digital technologies. Traditional carmakers are now using 3-D printers to create certain parts while leaving the rest of their mass-manufacturing process intact.

When it comes to speed, 3-D printing cannot compete with the astounding production of a traditional assembly plant. The question is whether the 3-D printing process will be fast enough for minifactories to make money in niche markets—or if it will be fast enough in three or four years.

DIVERGENT 3D'S FIRST car, which Czinger showcases at mobility conferences, is a sleek, purple sports car called the Blade. It has the curves of a 1950s-era Porsche, yet it weighs only 1,400 pounds. If it had a more modest motor, it could run more than one hundred miles on a gallon of gas. But the show version, built to wow motor-heads and car writers, sacrifices economy for performance. It has the power of 700 horses, and it can accelerate from zero to sixty in a blistering two seconds flat.

The Blade is only a concept car at this point. Divergent's business, Czinger says, will not be in producing and selling cars, but instead in leasing its manufacturing system—its software—to automakers big and small around the world. Even as 3-D printing grows faster, it will never compete with the productivity of mass manufacturing. But niche markets don't require such speed or scale. Czinger estimates that simple printed cars with commodity engines will sell for $6,000 or so—not even a third the price of Mark Frohnmayer's Arcimoto SRK. The plans for the Shanghai factory call for the production of nearly one thousand such cars per month. The number is minuscule by mass-manufacturing standards, but it's an entirely different business model. If this approach takes off, Divergent 3D could become a global software platform for vehicle manufacturing. That is Kevin Czinger's goal.

2

LA: Crawling to Topanga Canyon

The story of Los Angeles, like that of most cities, is a tale of changing mobility. Until the 1880s, LA was a small river town at the foot of the San Gabriel Mountains, fifteen miles from the Pacific shore. The few thousand European Americans who lived there had either made a harrowing transcontinental journey across mountains and deserts in wagons drawn by animals, or they'd taken extravagant detours in ships, some of them sailing twelve thousand miles around Tierra del Fuego.

When trains from the east finally reached the town, they unleashed a wave of migration. In 1888, as two competing railroads, Union Pacific and Santa Fe, battled for the nascent Southern California market, a price war erupted. A passenger in Kansas City could travel all the way to Los Angeles for as little as twenty dollars, then ten dollars, and during one spectacular promotion, a single dollar. To these first travelers, this rolling conquest of an entire continent must have seemed magical.

With the new railways, the West was tamed. Travelers simply stepped into a railroad car at Kansas City's new Union Depot. Two or three days later, they would find themselves far beyond the Rockies, the Sonoran Desert, and the Sierra Nevada, descending into a boomtown of palm trees and rolling surf, a city of angels.

What wasn't to like? Los Angeles grew, and a railroad heir named Henry Huntington avidly pushed it along. Huntington arrived in Southern California in the 1890s and promptly began to buy up streetcar companies. He extended tracks throughout the region. He also acquired an electricity company, which fueled his trains. Nearly a century before the word existed, Huntington was a master practitioner of synergy.

He was also a visionary. He was in fact building transit for an immense city that did not yet exist. While he ran the electricity and the trains, the biggest money, he knew, was in real estate. The history of railroads made it clear that when tracks came through, the value of nearby land soared, turning scrub lots into premium parcels. So he bought loads of acreage, from the foothills town of Glendale and Pasadena all the way to Redondo Beach, and then connected them to his train lines. People who settled in those areas could ride his Red or Yellow cars downtown for work and shopping.

It was a comfy racket for Huntington and a handful of other investors, but it was headed for a big disruption, similar in many ways to the one we're facing now. The first automobiles showed up early in the twentieth century, and after 1908, when the affordable Model Ts started rolling off Henry Ford's new assembly lines, Angelenos embraced them. In 1911, to meet soaring local demand, Ford even set up a Model T factory on Seventh Street, in downtown Los Angeles. By the following year, one in every eight adults in Los Angeles had a car, more than twice the rate in such established cities as New York. Some of these new automobiles began to compete directly with Huntington's streetcars. Jitney cabs trolled the streetcar lines, snatching away their passengers and collecting the same fee, a nickel. A century later, this same pattern is repeating in hundreds of cities, as commuters abandon buses and subways for car-sharing services.

As Angelenos took to the wheel, downtown LA grew congested. The streets effectively became narrower because of the parked cars. They were also more crowded, because drivers then—just like today—roamed in search of parking. The streetcars, which still served the majority of Angelenos, were getting tied up in traffic. As Richard Longstreth describes in his study of LA, *City Center to Regional Mall*, the city council addressed this problem by voting in 1920 to ban streetside parking in the urban core.

This raised vivid concerns. The *Los Angeles Times* warned that if people couldn't park downtown, other competing downtowns would spring up, in lots of other neighborhoods, including Hollywood and Glendale. Those competing downtowns would sprout their own theaters and department stores, and with their cheaper real estate, they'd be able to offer parking. This would hollow out LA's downtown.

Within a decade, the new order was firmly established. Cars prevailed, and again, it was real estate that drove the change. Developers built in areas far from public transit, because now people could drive. Those who moved into those suburbs were self-selected car people. During this decade of tumultuous expansion, LA fleshed out much of the land between the bare bones of the streetcar networks. In 1927, the director of the city's planning department, an auto buff named Gordon Whitnall, wrote: "So prevalent is the use of the automobile here that it might almost be said that Southern Californians have added wheels to their anatomy . . . that our population has become fluid."

The alarming point about this transition, from the perspective of Angelenos who don't want history to repeat itself, is that it just seemed to *happen*. From the perspective of everyday folk in the area, the process was simple: There was this new way to move. People liked it. And it promptly overwhelmed a large piece of the planet; in this case, Southern California.

Peter Marx, a professor at USC and a former chief technology officer for Los Angeles, goes so far as to suggest that the automobile in Southern California is the dominant species, and that we human beings serve it. This is tongue in cheek, of course, but his logic has the ring of truth.

Automobiles do seem to lord over us. We pay them extravagantly, and we polish and clean them—even though they tend to be lazy. Most of them labor for only a fraction of their existence, and they snooze the rest of the time in expensive lots and garages that we build for them.

Human beings in Los Angeles spend thousands of dollars every month to park their own bodies, many of them in cramped studios and duplexes. The city has an urgent homeless problem, with more than fifty thousand of its people living without a roof over their heads. The automotive overlords, by contrast, often park for free. Each one, on average, luxuriates in more than three spaces, many of them safer and more substantial than the ramshackle shelters and tents that homeless humans occupy. Parking spaces in greater Los Angeles take up some two hundred square miles. That's five times the entire footprint of Paris.

The whole environment has been molded to the needs of these four-wheeled beings. If aliens came to earth and studied our species, they might notice that we appear happier when at the beach, on fields, in the mountains, or surrounded by vegetation. But those beautiful environments are inhospitable to cars (though useful as backdrops to advertise them). So for a century, as Joni Mitchell sang, we've paved paradise and put up a parking lot.

To underscore our subservience to these machines, one more point: if, through carelessness or pride, we venture onto their paved terrain, they can kill or maim us.

SO IN THIS book about the coming age of mobility, why include a chapter about this iconic car town? The three other cities discussed in this book clearly make sense: Helsinki has its mobility apps; Dubai is spending billions for every new piece of technology on the shelf. Of course it would be silly to ignore China, so sure, Shanghai.

LA, though, doesn't seem to fit. LA County minted the global model for highway-connected sprawl. With its eighty-eight municipalities, the county is shaped by thousands of miles of highways, driveways, boulevards, parking lots, and culs-de-sac. This is yesterday's infrastructure. What does it have to do with tomorrow?

That's precisely the point. Los Angeles, more than traditional compact cities, like San Francisco or Paris, must reinvent itself. The city's leaders, including Mayor Eric Garcetti, have vowed to do just that. LA pioneered motorized mobility one hundred years ago, they say, and it can do it again. "My goal—and the goal of this city—[is] to be the transportation technology capital of the world," says Garcetti. The challenges, as we'll see, are immense. But the transition, when it comes, will likely be far more dramatic—for better or for worse—than in most other cities.

Consider the mileage. Drivers in LA County travel a combined average of 221 million miles a day. That's the equivalent, every day, of a round-trip to the sun plus a one-way jaunt to Mars. You could argue that Angelenos would travel far more than that, maybe twice or three times as much, if the road experience were less miserable. The average LA driver spends one hundred hours a year stuck in traffic. Some commuters suffer multiples of that.

So for electric bus manufacturers, designers of flying machines, ride-sharing app developers, tunneling companies—for mobility entrepreneurs of every stripe and color—LA represents an immense and ravenous market for miles. The coming transition is bound to transform the economy, the cityscape, and life itself in this sun-soaked expanse of California.

In many ways, LA's next step could be a return to its past. Christopher Hawthorne studies this history. For fourteen years, he was the renowned architecture critic for the *Los Angeles Times*, and more recently crossed the street to city hall, where he has a brand-new title: chief design officer. His job is to think about the emerging layout of the city: where people will live, work, study, and play, and—naturally—how they'll move from one place to another. This has led him to conclude that in the twentieth century, there were two very different LAs, and now, in this new century, we're seeing a third.

The first LA, Hawthorne says, started out as a traditional urban center, with a strong civic culture rooted in its downtown. That was the destination of Huntington's streetcar lines. It was where people worked and shopped and went to the theater. You can still see vestiges of this first LA, from the grandiose Union Station, built in Spanish villa style in 1939, to the thirty-two-floor art deco tower that crowns city hall.

The second LA, as Hawthorne sees it, took shape following the Second World War. Much as the *Los Angeles Times'* editorial writers had feared twenty-five years earlier, LA had extended into a policentric urban region with dozens of smaller downtowns, all of them connected by a fast-growing network of freeways. It was a vast sprawl.

For a time, it was a wildly successful one. Hollywood led the globe in entertainment. Fueled by Cold War spending, LA grew into a manufacturing giant and a world leader in aerospace. And LA's freeway culture was central to its brand, one of freedom, sunshine, and sex. "She makes me come alive," the Beach Boys sang, "and makes me want to drive."

Yet as Hawthorne sees it, the civic spirit, the sense of belonging to something in common, slackened during this period. This wasn't unique to LA, of course. But in Los Angeles, as is often the case, the change came earlier and was more extreme. People spent more time cocooned in their homes and their cars, and less of it within chatting distance of fellow Angelenos. With growing crime in the area, the most gruesome cases looped endlessly on TV, the streets themselves started to feel dangerous. In the second LA, the car was a safe space, a shield.

Now Hawthorne sees a third LA rising. In many ways, it recalls the original version, the one centered on a downtown, where people bumped into one another on electric trains and crowded sidewalks. In this process, LA doesn't revert to a single downtown. But it does become a denser place, with more people packed into many of its empty spaces, more of them living in apartments. This more concentrated population will depend far less on cars, and fewer of its inhabitants will need to own one.

This shift is already under way. You can see it in the Arts District, a fifteen-minute walk from city hall. A neighborhood of old warehouses and small manufacturing plants, it's now sprouting galleries, cafés, converted lofts, and new apartment buildings. People bike and scooter and stroll on sidewalks. This preview of the third LA looks and acts more like a traditional city.

For this filling-in trend to spread across greater LA, Angelenos will need a host of new mobility options. These extend from the familiar—walkable sidewalks, bike paths, new Metro lines—to technology's cutting edge: think autonomous air taxis and high-speed pods blazing through tunnels.

MAYOR ERIC GARCETTI, leaning back on a couch in his spacious city hall office, is reminiscing about his first car, a 1975 gas-guzzling Ford Torino station wagon. He reaches for his phone and does a fast image search. "Here it is," he says, pointing to a boxy behemoth with fake wood paneling on the sides. "It was more a question of gallons per mile than miles per gallon."

It was a couple of months after Garcetti's sixteenth birthday, in 1987, that his father, Gil (who would later serve as LA's district attorney), bought back the Torino from the man he'd sold it to and presented it to his son. Even with its miserable gas mileage, the Torino represented freedom, and driving in LA still seemed fun. Today, the mayor says, LA still has some "amazing drives," up the Pacific Coast Highway, for instance, or the twists and turns of Mulholland Drive. Topanga Canyon is still gorgeous. Add it all up, and the nice drives occupy "about two percent of the time," Garcetti says grimly. "The other ninety-eight percent you're in traffic." And crawling.

Driving in LA is mostly an exercise in tapping the brakes. Eastbound rush hour traffic trudges on the 10, from Santa Monica to downtown, at about nine miles per hour, no faster than a six year old on a bike. In fact, many routes in LA are slower today than they were in the 1920s, when people were still driving Model Ts.

Immobility saps the Los Angeles region of its very essence. The whole point of living and working in a city, after all, is to connect with other people. In LA, it might be to haggle over a merger in Pasadena, to play tennis on the courts at UCLA, to celebrate a *quinceañera* in Boyle Heights, to dream up a screenplay over drinks in Marina del Rey. People interacting with other people is akin to a city's nervous system. For it to work, people have to be able to move. Otherwise, why pay the rent to live in LA? You might as well FaceTime from somewhere else.

For decades, the answer has been to widen the highways—which is almost always an exercise in futility. Way back in 1955, the social critic Lewis Mumford quipped that adding highway lanes to ease congestion was like loosening your belt to cure obesity. More supply simply generates greater demand. The most recent example was a $1.1 billion widening of the key north–south artery, the 405. The job, which involved hacking out a wider pass through the Santa Monica Mountains, took four years. It led angry commuters through miles of convoluted detours. Once complete, that stretch of the 405 was as slow as ever. Tesla's founder, Elon Musk, an LA resident since 2002, laments to a gathering of Bel-Air residents that the 405, even with the improvements, "varies between the seventh and eighth levels of hell."

Absent this powerful and widespread frustration with LA's car monoculture, local authorities would have little hope of transforming mobility in the region. But even car-loving Angelenos now see that the second LA is unsustainable. In 2016, voters were so fed up with the status quo, they approved Measure M, which hiked gas taxes to finance $120 billion over four decades in transportation spending, much of it in Metro expansion.

For a target, Garcetti focuses on the LA Olympics in 2028. His office is decorated with enormous black-and-white photos of LA's two previous Olympics. The 1932 games, LA's coming-out party, left the city its iconic Coliseum. The Olympics of 1984, staged at the height of the Cold War, felt like a victory parade. First, LA beat out New York City for the games, which made it sweet. Then Americans, like the sunny gymnast Mary Lou

Retton and the sprinter Carl Lewis, scooped up loads of medals (thanks in part to the Soviet and Eastern Bloc boycott). In the '84 Olympics, LA even figured out a way to make a profit—which enhanced its reputation as a can-do region, one capable of economic miracles. *Time* magazine named the games' organizer, Peter Ueberroth, its 1984 Man of the Year.

By the time the 2028 Olympics roll around, in Garcetti's vision, the third LA should be on full display, the city's mobility largely transformed. County-controlled LA Metro plans to have twenty-eight major transit projects completed in time for the '28 games. They include a doubling of the Metro line, the introduction of electric buses, the availability of subsidized electric rideshare services for the poor and disabled, and the expansion of bike lanes and pedestrian greenways.

The city and county are wooing mobility players big and small, whether they're running fleets of autonomous jitneys, operating air taxis, or building electric buses. "If you think about the finance capitals of the world," Garcetti says, "New York and London come to mind. Car cities, Detroit and Munich, Tokyo. But what city is the leader in transportation technology? That's what I want LA to be." He points to the city's leadership in aerospace, the five-hundred-odd tech start-ups along the western strip (so-called Silicon Beach), the two Hyperloop companies, and an openness to experimentation. "I want everyone to come here and try stuff," he says. "I want LA to be the kitchen where this is all cooked."

At the same time, he sees LA's paved roadway, the asphalt tapestry smothering much of the region, as an asset. The roads can be repurposed, he says, parts of them transformed into bike paths and walkways. "Look at the High Line," he says, referring to a converted stretch of elevated railway in New York City that is now a world-famous park.

Of course, many of the roads will remain focused on their current job: the movement of automobiles. Even as other options emerge, the automobile isn't deserting Los Angeles any time soon. The region has 6.4 million cars and trucks, and the average vehicle stays in operation for eleven years, many for much longer. So on the eve of the 2028 Olympics, millions of cars would be still circulating in LA even if Angelenos, en masse, stopped buying them today. And that's not happening.

The idea is not to get rid of the car, but to end the car monoculture. It's a matter of giving people choices. As this happens, the city will become multimodal and greener. That part of the LA vision sounds like Helsinki, Shanghai, and hundreds of other cities around the world.

In LA, though, one problem stands out: while the city and county are spending billions on new Metro lines and expanded bus service, it's not yet catching on. Only 7 percent of the population rides public transit. Worse, these numbers have been *shrinking.* Growing numbers of working poor in the area, data show, are ditching the buses and trains, and instead buying used cars.

This is the downside of economic success. Since the rise of the automobile, public transit in LA, as in much of the car-centric world, has largely served the underclass, including many who cannot afford their own personal vehicle. But now, just as the city leaders try to transition away from the car economy, growing numbers of Angelenos are "graduating" into it. They add to congestion. So do the thousands of Uber and Lyft drivers plying the streets and byways in search of passengers. After decades of progress against smog, LA's air grew dirtier in 2016 and 2017, registering some of the nation's highest ozone counts.

Garcetti boils the problem down to one of geometry. You have millions of people trying to get from point A to point B, he says, "and they're all occupying the same plane." That plane is defined by the surface of the earth and the roads plastered on top of it.

To visualize the limits of the status quo, think of all the people working in the city hall building. Since they're organized in layers—thirty-two floors—they're not too crowded. But most of those people came to work—at rush hour, no less—smushed onto a single plane, along the same ribbons of highway. Only later do they ease this crush by stacking vertically. One solution to the congestion, Garcetti suggests, is to add new planes for mobility, above the earth and beneath it.

FOR SELETA REYNOLDS, urban aerial mobility is not a current pressing concern. Nor are the tunnels Elon Musk's Boring Company has begun to dig (which we'll visit in chapter 6). Reynolds heads the LA Department of Transportation, and her focus is on moving millions of people, most of them by traditional means. She thinks much more about providing decent transportation to have-nots, and doesn't worry much about the people crawling in Lamborghinis or Porsches up and down the 101.

Reynolds, a Mississippian with a history degree from Brown, came to LA from San Francisco, where she'd been an activist manager in the Municipal Transportation Agency. Her division was Livable Streets, and she oversaw the launch of Bay Area Bikeshare. In 2014, Mayor Garcetti brought her to Los Angeles as the top transportation official. She now sees the city at a crucial transition, much like the one a century ago, from streetcars to automobiles. Back then, for-profit corporations with bottomless reservoirs of dollars for promotion and greasing of the political levers sold the public on an enticing technology. And once the public was behind the wheel, it was hooked. By that point, the government was reduced to satisfying motorists' demands:

building highways, making sure people had places to park—in short, surrendering to the car.

All these years later, as Reynolds sees it, LA has a second chance. Tech entrepreneurs are promising a transformation—mobility that's cheap, green, and fun. But from Reynolds's perspective, if the government doesn't step in and assert a measure of control over this new ecosystem of mobility, the technologists will dictate the shape and nature of the city for the coming century. In that case, it will be made to fit their needs, and their bottom lines, and not those of the eighteen million people living in greater Los Angeles. "We could repeat the mistake we made a century ago," she says. "We adjusted the city to the technology, instead of the other way around."

Reynolds focuses most on social equity. Entire neighborhoods of LA are transit deserts. She cites the example of Boyle Heights, a historic Mexican American enclave near downtown that is fenced in by freeways. "If you live there without a car," she says, "you're trapped." Those who cannot move across the region efficiently and at an affordable price lose out on opportunities. They struggle to get to trade schools or to job interviews. Entire job markets lie out of reach. This trend worsens with time. The poor migrate to neighborhoods where the rent is cheap, which often are affordable, in great part, because the transportation is so terrible. In these mobility deserts, the poor get poorer, and many fall behind on their rent. This in turn feeds the epidemic of homelessness in LA.

The obvious answer is to wean Angelenos from their addiction to cars. LA Metro is spearheading this strategy by dramatically extending its network. Plans call for new lines stretching to the airport, and others linking downtown to Hollywood and the UCLA campus in Westwood. By the time the Olympic torch is lit in the LA Memorial Coliseum, according to the plan, LA

County may boast the second-largest urban train system in the United States, behind only New York's.

To some evangelists of the mobility revolution, this seems a bit retrograde. Who will take the Metro if Elon Musk's underground network springs into action, zipping around Angelenos like packets of data? And what about the fleets of autonomous pods, the next generation of dockless electric bikes, the air taxis?

Like Reynolds, Phil Washington, CEO of LA Metro, sees public transportation serving as the main trunks and branches of the third LA, with the scooters, bikes, and car-shares working as connecting tendrils. His logic is based on the immense distance to be traveled—all those round-trips to the sun. Millions of people, he says, must share a good portion of those miles. Otherwise, if each trip carries a single person on a single trajectory, the next stage of transit could be knotted up even worse than rush hour on the 405 at Sepulveda Pass.

From Washington's point of view, the competition between public transit and all the mobility entrepreneurs just isn't fair. LA County's Metro, like other public agencies, has a mission to provide service to everyone. This dramatically raises costs. In the United States, for example, the Americans with Disabilities Act of 1990 mandates universal access to public transit. Metro must spend for buses that hoist wheelchairs from the curb, and equip train stations with ramps and elevators. It must also provide transport to isolated neighborhoods, and look out for those lacking a smartphone to hail an Uber. This is expensive. It serves the people who need help—whom mobility entrepreneurs can blithely ignore. Those new players, Reynolds says, "are building businesses on our infrastructure. And we get nothing in return."

What's worse is that if new privately owned services take off willy-nilly, each one rushing to develop its own niche, they'll claw for market share by selling mobility at bargain-basement prices. A century ago, when this happened, the government was ineffectual. It ended up being an enabler of the automobile monoculture. Those cars also had the right, at least for their first century, to pollute, congest, and roar. Running over people was collateral damage.

Like most city officials, Reynolds is eager to avoid a reprise of that mess. She wants not just to supervise traffic, but to manage it. This isn't easy, because broader LA County is fractured into scores of fairly independent municipalities. And Metro, like most transit agencies in the United States, is not particularly popular. Most motorists use it rarely (while wishing others would).

Yet in Los Angeles, as in cities worldwide, someone, some entity, is going to be monitoring and managing our movements. This will happen more in some places than in others. But this control is the nature, and the promise, of connected mobility, the key to cleaner, safer, cheaper, and faster movement. The political fights ahead will focus not just on the extent of this control, but also on who's in charge. Governments are sure to jostle with businesses large and small, and the crucial factor will be data.

In coming years, practically every conveyance will be linked to networks. Most already are. But in Los Angeles, as we'll see in Helsinki, the data travel in different streams. The subway counts its riders, as do dockless bike companies like Lime. The phone companies can track the movements of their users, as can Google and Facebook. When autonomous cars hit the roads in growing numbers, each one will gush thick rivers of data.

Yet at this point, no one has access to the entire span of mobility data. This is perhaps the ultimate treasure in the coming era. Whoever controls the data will be in a position to manage movement, and to build businesses on it. LA took a first step in 2019, establishing data specifications for bike-share and scooter services. This will allow officials to track their coverage, and coordinate with public transit. This common data standard is now dubbed MDS, for mobility datasets.

The key to controlling networked mobility is data. But who will own it? Could one dominant player, perhaps a company like Google, become the de facto mobility platform?

The strategy in the LA government is for the city to control the data, and to share it with other players on a need-to-know basis. From Reynolds's point of view, who else can claim to be looking out for everyone?

Transit officials in LA have in fact been working on optimizing traffic for decades. Take an elevator down four levels beneath city hall, pass through a series of security barriers, and you'll find yourself facing walls of TV screens in the Automated Traffic Surveillance and Control Room. Set up for the 1984 Olympics and originally funded by the federal government, it provides video feeds on hundreds of intersections and problem spots in LA. Back in 1984, the data analysis was primitive: human beings looked at TVs and saw traffic jams. The tools they had to respond were also unrefined. They could lengthen red or green lights where they saw problems, or in serious cases dispatch a traffic cop to the scene.

In the modern version, the surveillance network has spread to 4,700 intersections. In addition to images, sensors send in detailed reports on the traffic conditions: the number of vehicles passing, their speed, the flow of pedestrians. An AI engine churns through this information and attempts to optimize the flow, with a priority for mass transit. If a Metro bus, for example, is running behind schedule through East Hollywood, the computers can automatically orchestrate the lights on North Vermont Avenue to hurry it along.

Over time, the system will attempt to orchestrate the movement of growing numbers of vehicles, public and private alike. Autonomous cars, for example, will be sent along the most efficient paths and rerouted at a moment's notice. Traffic lights will flash routing instructions to other cars, as well as to cyclists, and even to pedestrians.

When it comes to surveillance and control, as we'll see in Dubai and Shanghai, this is still kids' stuff. But unlike those cities, where government control has long been a fact of life, LA is attempting to manage millions of people who associate driving with freedom, who often view government with suspicion, if not contempt—and who can rebel at the polls against officials who rub them the wrong way.

ON THE BUSY pedestrian walkway in the oceanside municipality of Santa Monica, a black electric scooter, a Bird, is leaned against a palm tree. The Bird company, a hometown start-up, launched its business one autumn night simply by littering hundreds of its scooters throughout the eight square miles of Santa Monica, from its Ferris-wheeled pier all the way to the Brentwood town line. Residents who stopped the next morning to inspect these two-wheeled apparitions noticed a link to a smartphone app. Many downloaded the Bird app that very morning and started activating the scooters with their phones. (As a condition in the rental agreement, they vowed to wear bicycle helmets, but residents broke that promise en masse.) Paying a dollar for each ride, plus fifteen cents a minute, they were zipping around town with very few questions asked, other than "Will it get me there?" The answer was a resounding yes.

The day after the Bird scooters descended on Santa Monica, the company's founder, a round-faced young man named Travis VanderZanden, reached out on LinkedIn to the mayor of Santa Monica, Ted Winterer, according to an account in the

Washington Post. He invited him to drop by Bird's headquarters. It was just a few blocks away. There they could discuss Bird's mobility strategy for the region.

First, the mayor told him, they had a few legal issues to discuss. After all, Bird had never asked for authorization, much less discussed safety issues, sidewalks, or parking with the city. The company hadn't said boo.

It didn't take long to recognize a familiar pattern. VanderZanden, it turned out, was a former executive at both of the ride-sharing powerhouses, Lyft, and more significantly, Uber. The practice at Uber was to launch the service first and hammer out the legal details later. This had an impeccable logic. Once a mobility start-up gained lots of riders, it could attract millions of dollars from venture funders. Then it could use that money to hire armies of lawyers and make peace with the cities—to get legal. The key was to come to these negotiations as a popular and newly rich powerhouse, and not a supplicant.

The day that Bird launched in Santa Monica, VanderZanden was still bankrolling the company with just a small round of angel funding. But that changed quickly. Within weeks he had attracted $15 million in VC money, and another $100 million soon after that. By early 2018, when the company agreed to pay a $300,000 fine to the city, its valuation topped $1 billion. By that point, the Santa Monica fine was something Travis VanderZanden could have jammed into a tip jar. He had an enthusiastic customer base, a plan to expand nationally, and a fearsome fortune. Cities would take him on at their own peril.

This model extends throughout the tech economy. Giants like Amazon, Google, and Facebook have behaved like jumbo versions of Bird. They start by providing valuable services to billions, at low cost or for free. Largely unencumbered by government regulations in their early years, they go on to build market capitalizations bigger than entire industrial economies. In a Vice News feature, Euwyn Poon, cofounder of the San Francisco scooter company Spin, called this "innovating on the regulatory side."

What's more, because of the services they provide, these tech companies are often more popular than governments, more trusted, and widely viewed as essential. It's often when something goes wrong—when their autonomous machines kill someone, or a tech platform is used to spread lies or hack an election—that governments can start to regulate them. But by that point, the tech companies can fight back, with their legions of customers, advertising might, and armies of lobbyists.

This build-first, fix-later model could spell catastrophe in the mobility revolution, and especially in car towns like LA. Seleta Reynolds envisions autonomous pods, pimped out as rolling lounges, carrying people on joyrides at low prices. These pods could be small bars or restaurants, game pods with virtual reality hookups, even smoking chambers offering rich selections of California's legal marijuana. The skies could be black with drones, some of them flying across the Santa Monica Mountains just to pick up tacos or a roll of toilet paper.

If governments fail to assert their control with taxes and regulations, cheap, ubiquitous mobility services could overwhelm the entire region, much the way the automobile did.

STILL, AFTER A century of the automobile monoculture, the region's mobility revolution also offers opportunities to tackle some of LA's toughest challenges, including homelessness. According to Zillow, the real estate database company, a median-earning family in Southern California must pay a staggering 46.7 percent of its income to rent median-valued housing. This is the highest such rate in the country. Some 130,000 households can no longer afford this expense and are transient or homeless. The most vivid display of this scourge is the sprawling LA tent city known as Skid Row. It occupies a square mile between downtown and the Arts District and is home to some ten thousand people in wretched conditions.

As we mentioned earlier, the physical structure of LA favors cars, in many ways, over people. But as Angelenos find other ways to move around, many of them will be left with a legacy of the automobile age: the garage. Steven Dietz, a leading venture capitalist based in Santa Monica, recently launched a start-up called United Dwelling. The business plans to convert unused garages into affordable housing. With the right incentives, he says, this could dramatically expand the housing supply within LA, reducing its century-old sprawl.

It doesn't even have to wait for the air taxis, speeding underground trains, or other wonders of the coming age of mobility. Not long ago, Dietz, who also teaches at the University of Southern California, had his students knock on doors in LA to find out what people kept in their garages. In a study of seven hundred residences, all of them with two-car garages, only 8 percent of homeowners kept their cars in the garage. "The rest of them parked in the driveway and used the garage to store their stuff," he says, laughing.

"This could work," says Mayor Garcetti, back at his office. He does some quick math. "We have five hundred thousand single-family homes in LA," he says. "If ten percent of them could convert their garages, that would be fifty thousand new housing units."

The real estate opportunity extends far beyond family garages. The Los Angeles International Airport, known as LAX, is currently building a parking garage with 4,500 badly needed spaces. But as travelers find different ways to reach the airport, from Metro and ride-shares to flying taxis, those parking spaces might go begging. With an eye to that shift, says Justin Erbacci, chief innovation officer at LAX, the new parking lot is designed to be converted to either retail space or housing. "We're making flat floors and higher ceilings," he says. Instead of building the ramps inside the building, Erbacci's team placed them on the outside, where the erstwhile garage, when the time comes, can shed them.

This is what the mobility revolution has in store for LA. It's a real estate story, as it always has been. The value of each piece of land, and the use for it, depends largely on how people can get there, and where they can go. This was the case in the era of streetcars, and again in the century dominated by the automobile. Once more, changing mobility is poised to reshape life and geography in Southern California.

3

800 Electric Horses

On an early summer day in 1896, the neighbors on Bagley Street in Detroit heard an awful banging. It came from a rented garage, where a thirty-two-year-old engineer named Henry Ford had pieced together a new creation, a gasoline-powered "quadricycle." To his great frustration, he couldn't fit it out the door. So Ford had grabbed an ax and was pounding away at the brick wall to widen the exit for his horseless carriage.

The racket from Ford's garage announced a new age of human mobility. But from a manufacturing point of view, the auto industry as we know it wasn't born until 1908, when Ford's first assembly line, at nearby Highland Park, launched the mass production of the Model T. That event marked a leap in the development of manufacturing as an engineering science. Ever since then, the global auto industry has been relentlessly fine-tuning Ford's original process, wringing out inefficiencies while producing millions of extremely complex machines at prices billions can afford. It's no mean feat—and one that put its stamp on virtually every aspect of human society in the twentieth century.

The automobile industry is a triumph of human organization—and, at the same time, a template for the advances that await us. Car companies have perfected ruthlessly efficient and repetitive processes, breaking down the immense complexities within them and optimizing every necessary step and movement. Chaos may have reigned outside the factory walls, in the streets and schools—but the auto manufacturing plants came to be microcosms of order.

Now the synchronized standard of a factory is spreading into our lives. To understand why, picture one of Ford's early engineers, stopwatch in one hand, clipboard in the other, conducting an efficiency audit of human mobility in an American neighborhood, circa 2020. In most driveways or garages he spots a parked car, maybe two or three of them. Scandalous idle capacity. A pickup truck big enough to haul a heifer or two pulls into a 7-Eleven, and the driver emerges a minute or two later carrying nothing but a steaming cup of coffee. More waste. Jumbo buses rumble down boulevards carrying only a handful of passengers. Even at deserted intersections, they stop at red lights, burning fuel and time. A factory as wasteful as this would be shuttered within a week.

For the first century of the automobile age, the prodigious production inside a factory and the immense waste outside it made for a perfect match. The more wasteful we were, the more we consumed, and that demand kept the factories humming. But now we're turning that century-old status quo on its head. Indeed, many of the greatest opportunities in the mobility revolution come from slashing waste. "If you think about what we're building," says RJ Scaringe, an automaker we'll meet in a page or two, "it's the insertion of efficiency into transportation." And for this, the methods of the factory are extending to the rest of the world.

It all revolves around data. This has long been the case in factories. The foundation of industrial management involves knowing where everything is—every component, every worker, every machine—and then mapping out each one's schedules and movements. In the early days, workers marked down this data on handwritten ledgers. In today's manufacturing, rivers of digital data flow second-by-second into powerful analytics programs. Feeding continuously on this data, the software optimizes every movement in a factory. That's its job.

Ubiquitous data is starting to power an analogous process in the mobility economy. We enable this process by communicating our locations and destinations, mostly with our cell phones. As thousands of new sensors permeate the coming generation of cars, buses, highways, and traffic lights, today's streams of data will grow into raging rivers of it. Increasingly, just as in factories, this data will feed optimization engines. In time we'll find ourselves shuttled along the most efficient paths, much like the fenders and ball bearings that make their way from distant suppliers to the assembly lines piecing together Toyotas or Ram Trucks.

This economy of movement promises enormous gains in efficiency. The paradox is that while Fordism enabled mass production, similar efficiencies in our society will *decrease* demand for cars and trucks. We'll be able to get from one place to another with fewer of them.

This isn't to say that cars will go away. But we'll make better use of them. Growing numbers of us will share them, subscribe to them, rent them, perhaps even fly in them—but many will avoid buying them. This will have a profound effect on the automobile industry, an essential pillar of our industrial economies.

For auto manufacturers, the math is all too simple. Say we manage to increase the average usage for each car from the current 5 percent to a still pathetic 7 percent. How does this happen? In the most elementary case, two neighboring families agree to carpool. One of them decides it needs only one car, not two. If enough people come to that conclusion, whether spurred by carpooling, autonomous ride-sharing, or a spate of new scooters, it could render a big chunk of the global auto fleet superfluous. A study by PwC estimates that Europe and the United States will shed a quarter of their auto fleets, a combined 138 million cars, by 2030. The same study projects growth of 100 million of them in China.

This leads to one central question for this chapter: Why would anyone want to be launching a new car company in Europe or the United States in the 2020s? It's counterintuitive. That much we can say.

ONE OF THE first things to appreciate among the rarified ranks of twenty-first-century automobile entrepreneurs is their gift for delayed gratification. A smart kid can create a phone app in her dorm room and mount a business on it within weeks. Building a new car company is a far more deliberate venture, torturously so, and for most, prohibitively expensive.

Just ask RJ Scaringe. The Floridian is just a year older than Facebook's founder, Mark Zuckerberg. But their career paths could hardly be more different. Since his college days, Zuckerberg has built his giant digital enterprise, raked in billions, and weathered storms over privacy and rigged elections. His early career was the subject of an Aaron Sorkin biopic, and that movie is already *old*.

Scaringe, tethered to the physical world, is just getting started. He has been running his new electric car company, now called Rivian, since Barack Obama's first term in the White House. On this summer morning in 2018, he has yet to earn his first dollar of revenue, much less see the first Rivian roll off an assembly line. He scrimped in the early years, but has lately been spending lots of money. At a sprawling headquarters in Plymouth, Michigan, Scaringe employs teams of designers and programmers and engineers. He's also staffing up a massive assembly plant in Normal, Illinois. In California, he employs a high-priced software team. All this activity and spending will have been going for a decade by 2020, when Rivian hopes to sell its first electric SUV or pickup truck.

That market debut might be Rivian's only shot. In the auto industry, where a single line of cars can cost hundreds of millions of dollars, start-ups are rare. If they flop once, they rarely get funding for a second chance. In Silicon Valley, they say, you get rewarded for failing. Not in the car industry. So 2020, for Rivian, is shaping up to be a bet-the-company year.

Tall, quiet, with Mediterranean features and dark-rimmed glasses, Scaringe has dreamed since his childhood of building his own car company. Against formidable odds, he has succeeded. The reward for it, though, is to labor in an industry that devours cash and delivers a steady stream of near-death experiences. And that's in the good times. Reilly Brennan, the CEO of the San Francisco mobility venture firm Trucks, sees the new auto entrepreneurs, from Scaringe to Elon Musk, as "dancing right at the edge of extinction." While the bold and irrepressible Musk may be this generation's media star, he adds, the measured, soft-spoken Scaringe "is the one you'd want your daughter to marry."

On a recent summer morning, Scaringe winds his way through his Michigan headquarters, past hulking prototypes of electric trucks and SUVs and a wall covered with artists' renditions of autonomous pods. He's discussing the thriving global auto industry. Despite all the buzz about the mobility revolution and a coming shakeout, the industry is raking in profits and setting production records. The vast majority of the new vehicles are powered by gasoline. Henry Ford would look down on the industry and smile.

Scaringe pulls out his phone and does a search for data on horses. "When do you think we hit the top of the horse economy?" he asks. A logical answer might be 1910, when Ford started mass production of the Model T, or maybe five years later, when the roar of internal combustion engines and the *beep-beep* of car horns were changing urban soundscapes. With the exception of far-flung farmers and stubborn equestrians, who in 1915 would opt for a horse?

Scaringe finds his answer on a site called Cowboy.way, and points triumphantly to a graph on the screen. In 1910, the horse and mule population of the United States stood at just over twenty-four million. In the following decade, even as the infant auto industry grew, the horse industry continued its own expansion, reaching its high of twenty-six million in 1920. "That," says Scaringe, "was peak horse."

His point is that the new rarely supplants the old overnight. Legacy technologies are supported by habits, infrastructure, and economics. A Model T in 1920 cost $550, about the equivalent of $8,000 today. Horses were a lot cheaper. What's more, people loved horses, just as many love traditional cars today.

So for more than a decade, and far longer on farms, a gradual transition took place. As the old and the new commingled, horse farms did record business. Like them, the auto giants a century later rake in profits while a host of electric, networked, and autonomous upstarts, from Musk's Tesla and Alphabet's Waymo to Scaringe's Rivian, are positioning themselves to supplant them. "This is peak big auto," Scaringe says—the last party before the painful shakeout.

But why shouldn't established automakers triumph in the next stage of mobility, creating the electric, networked, and, increasingly, autonomous vehicles (AVs) that will zip us from one place to another? No one can match the legacy automakers' manufacturing muscle. While new technologies, like Kevin Czinger's Divergent 3D, might power niche producers around the world, it stands to reason that to transport a global market of eight billion people, many of them still dependent on cars and others even smitten by them, mass production will have its role.

The key questions, Scaringe says, aren't whether there will be a place for mass-produced vehicles, but rather how many we will we need and how much we will pay for them. A smaller and more miserly market, he predicts, will punish the giants and open the way for newcomers like Rivian.

To picture the Darwinian frenzy ahead in the auto industry, consider this: in China alone, as we write, hundreds of start-ups, a few of them much bigger than Rivian, are out to produce new electric cars. Even if only a fraction survive, perhaps another dozen manufacturers will be probing for market niches while battling on price, performance, and battery life—and all this while Western auto markets could be shrinking. Scaringe's chosen industry is heading into a fitful time.

Back in a glass-walled conference room, Scaringe draws a bar chart on the whiteboard to illustrate the dynamics. The tallest bar looks like one of the new pencil-thin skyscrapers in New York. If it has one hundred floors, he fills it four-fifths of the way up with a neat diagonal grid. The entire bar, he says, represents the average car owner's spending on transportation. For the last century, the lion's share of it—the grid in his whiteboard rendering—has gone to the manufacturer. In fact, most of us pay so much to Ford or Volkswagen or Toyota that we have to take out *loans*.

The auto giants hog the transportation dollar, leaving only

pennies for companies that provide car services, outfits like Midas or Europcar. The biggest car company, for example, Toyota, has annual sales of $260 billion. Uber, for all the disruption it stirs up, feeds on relative crumbs, only $7.4 billion.

In the coming years, service companies are positioned to gobble up a growing share of the transportation dollar, Scaringe believes. He draws a smaller bar, representing future spending on cars, and fills it only halfway up with the diagonal lines. His point is that we'll spend less on automobile transportation, and a smaller share of it will underwrite the manufacturers. This is because fewer people will own vehicles, just the way fewer of us, in the age of Netflix and Spotify, buy our own movies or music. Mobility, increasingly, will be delivered as a service. "If you're in the business of selling assets, your business will shrink," Scaringe says. "If you're providing operations and using data, your business will grow."

To illustrate this future, he suggests a thought experiment. Imagine, he says, that you're faced with a torturous choice: You have to select a single set of clothes—one outfit—to wear every day for the next four years. No other options. What do you pick?

Well, if you want to move around, maybe to mow the lawn or play tennis, the outfit should be loose fitting, casual, durable. If it's hot out, long sleeves might feel stifling. At the same time, you might not want to wear a rugged hoodie to work, much less to escort your daughter down the aisle on her wedding day. For most people in the world's consuming classes, putting on the same set of clothes every day and for every occasion is a ludicrous idea.

Yet that's precisely what we do with cars. Those expensive machines parked in our driveways or garages represent a single solution for mobility. Ownership sticks the vehicle with us for years. The great majority of us use the same contraption whether to pick up a pint of ice cream or fifty-pound sacks of fertilizer, or even to take a road trip to Yellowstone or Prague. It's one size fits all.

In the near future, according to Scaringe, each one of us will have the equivalent of an entire wardrobe of mobility options, each of them suited to a particular occasion. For a drive from Paris to Madrid, for example, we might want a comfortable car with reclining seats, cruise control, a TV screen in back for the kids, and at least the rudiments of autonomy—enough to keep us on track if we nod off somewhere in the Pyrenees. That's why so many people buy big SUVs, especially in economies where gas is cheap. But when we take a ride-share across town, our needs are much different. Timing and price rule, and comfort barely matters. Scaringe thinks about ride-sharing for a moment, and he adds a third constraint. "You don't want it to *smell* bad," he says.

Eventually, he argues, we'll have access to multifarious collections of vehicles, and to the right one when we need it. The emergence of ride-sharing is but an early step in this progression toward choice—and for many, away from ownership. This change to subscriptions, he says—from a single suit for mobility to an entire closet of them—will push us to reinvent the entire auto industry, as well as its business model.

This promises pain for manufacturers. First, fleet operators and an entirely new ecosystem of service companies will claim more of the auto industry's overall revenue. The auto giants, recognizing this, are positioning themselves to run fleets, which eventually will contain autonomous vehicles. But no matter which companies run the fleets, they're sure to operate the assets far more efficiently than we do as car owners. If a fleet runs vehicles at a modest average of eight hours per day, that's still more than six times the rate of the average motorist.

Remember the imaginary efficiency engineer, with the stopwatch and clipboard, who was so scandalized by all the idle capacity in our driveways? He'll see a lot less of it as the service industry rises. This heightened efficiency spells lower sales for the auto giants—or less mass for the mass producers. That could result in an industrial bloodbath.

Let's leave the mass manufacturers, for now, to sort out their own problems. (We'll revisit them in the next chapter.) Now that RJ Scaringe has outlined the coming battle, the more relevant question is how *he* plans to profit from it.

RIVIAN'S HEADQUARTERS, IN an industrial suburb of Detroit, sits about halfway from the city's center to the university town of Ann Arbor. The sprawling facility there is the size of a Walmart Supercenter, with a jumbo parking lot to match. Most of the real estate inside looks like an industrial design studio, with mockups of electric cars and trucks scattered about.

In the executive niche of this facility, Scaringe leans back in his chair and tells his story. Listening to him, you'd think he was obsessed with money. The theme comes up again and again. It hangs on his every question. But it makes sense. When your goal is to build a new car company, money is like food or oxygen. Money spells possibility and sustenance. Even for a niche auto manufacturer, financing is measured in the billions.

On this summer day in 2018, Scaringe has half of one, or $500 million, most of it from a Saudi conglomerate, Abdul Latif Jameel. But it's enough, barely, to launch his own brand of electric car and—if he's lucky—join Henry Ford, Ferdinand Porsche, and a handful of other survivors in a remorseless industry.

The dream of building cars started, Scaringe says, when he was ten years old, in 1993. One day he learned that a neighbor in Cocoa Beach, Florida, was refurbishing classic Porsches. Scaringe stuck his head in the garage and ended up spending endless afternoons there, soaking up everything he could learn about cars.

Like so many others in mobility start-ups, including Kevin Czinger at Divergent 3D, Scaringe grew up as a motor-head—an extremely brainy and ambitious one. His love of cars isn't like the "passions" you read about in LinkedIn profiles, that dutiful ardor for human relations or customer resource management. The passion millions of humans experience for cars is real and visceral. Scaringe has felt that way since he first set foot in his neighbor's garage. (He eventually would name his car company Rivian by juxtaposing syllables from the Indian River, which runs near Cocoa Beach.)

After emerging from high school, he says, his goal was both simple and enduring. It was to build the next generation of automobiles, smarter and more efficient machines, better suited for a crowded and polluted planet. For this he'd need a company. For the company, he'd need money. To get that money, he needed the knowledge and pedigree derived from higher learning. He expresses it with the step-by-step logic of an engineer. "I focused hard on school to give myself options," he says. "I wanted to come up on the learning curve and enable myself to raise capital." He chose the Rensselaer Polytechnic Institute, in Troy, New York.

It was during Scaringe's college years that the public Internet exploded. Companies like AOL and Yahoo, and hundreds of dot-coms, were all swimming in investors' money. Naturally, lots of his classmates veered toward the Internet and the sizzling field of software engineering. But Scaringe stuck with cars. At a time when the digital economy promised gold, transportation's moving molecules seemed a bit retro, dirty, and boring.

Nonetheless, the car industry was full of opportunity and

research funding. It was still a $2 trillion industry. When he applied to grad schools, he had a handful of offers, but he ended up opting for a PhD program at the Massachusetts Institute of Technology, in the Sloan Automotive Laboratory.

For someone looking to build the next generation of cars, it's almost impossible to imagine a better gig than what Scaringe found for himself, at the age of twenty-one, in Cambridge. Through the Sloan Laboratory, the auto manufacturers *themselves* sponsored his research on the future of their industry. They paid him to study them and whatever lay beyond.

What he could see, ever more clearly, were rising ranks of disruptive digital technologies (some of them developed by his former MIT classmates). They were primed to reshape the industry. In the previous twenty years, the computer economy had taken over phones and cameras, stormed into media and retail, and laid waste to advertising. Transportation was its next target. For the reigning auto giants, with their massive plants, armies of workers, and ranks of bureaucrats, it shaped up as a mortal struggle.

This was the time for a smaller and more agile breed of carmaker, Scaringe thought, one unburdened by thick layers of legacy. He considered leaving the fellowship and launching a car company. It was 2006, the height of the real estate boom. Billions of dollars, many of them conjured up by fraudulent mortgage-backed securities, were seeding new technologies. Transportation was no exception. Toyota's first electric hybrid, the Prius, was a consumer hit, and a host of venture money was pouring into new electric automakers, including Kevin Czinger's ill-fated Coda.

Money was easy, and Scaringe, with his persistent dream of launching a car company, worried he might miss his chance. He pieced together a PowerPoint presentation for an automotive start-up, and he presented it at a couple of business plan competitions. Investors responded by pulling out their wallets. Some were offering him as much as $20 million to launch an e-car start-up.

Still, Scaringe held back. Twenty million, no doubt, would give him a good start. It was enough to build a small team and design a car. But after that, he would be faced with raising money again, serious money, enough to put a car into production. Who had a better chance of raising, say, a billion dollars: a PhD from MIT, or a guy who'd dropped out of grad school? He stuck around, kept climbing up that learning curve, and got his doctorate.

Scaringe says his original idea was to proceed step-by-step, perhaps working a decade in the auto industry before launching his own venture. However, things changed in 2008, when the bottom fell out of the global economy. The financial crash pushed the American auto industry to the brink of collapse. Sales dive-bombed. General Motors and Chrysler both went bankrupt and looked to the US government for life support.

This was the hardest of times to raise money to build a car. The happy money that had been flying around in 2006 was long gone. Worse, it hadn't just evaporated. Much of it was transformed into debt, which was weighing down potential lenders around the world. But at the same time, the vulnerability of the auto industry was no longer in question. Might this be the moment for a changing of the guard, a turn to twenty-first-century methods—smaller, leaner, greener, cheaper? That was the pitch Scaringe would make.

He made a case for a small and focused car company, and for a new business model. The company would survive on a smaller share of the manufacturing revenue, and it would add revenue by providing services, including customized software apps and updates. Scaringe would create the digital markets, taking cues from the likes of Apple and Amazon. The automobile, after all, was a networked electronic device.

If a small and resourceful mammal, sixty-some million years ago, could have laid out its case for how it would prevail in a world dominated by towering reptiles, its spiel would have sounded much like Scaringe's.

Even at the depths of the recession, Scaringe was able to scrounge up enough capital to hire a twenty-person team and design a next-generation prototype. It was a rugged luxury car. But most of the development took place on the car's innards—its battery and power train. The idea, which Scaringe would return to, was to sell this platform to other companies. They could build their own vehicles on it.

As Scaringe hunted for investors, he probed the funding network at his alma mater, MIT. That put him in touch with Mohammed Abdul Latif Jameel, a billionaire who had graduated with a civil engineering degree from MIT in 1978. In 1945, his father, Abdul Latif, had started out in Jeddah with a single gas station. Ten years later, he imported four Land Cruisers from a struggling Japanese auto company, Toyota. Arabia's desert proved to be an ideal terrain for the rugged Toyotas. Abdul Latif imported more. Eventually, Abdul Latif Jameel became one of the largest Toyota distributors in the world. After Mohammed took over, following his father's death in 1993, the privately held company branched into marketing, manufacturing, power generation, and financial services, with operations in thirty-one countries.

Through the MIT alumni network, Mohammed found out more about Scaringe and the prototype he and his team had developed. He became excited, and suggested targeting the surging middle class extending across Asia, the Middle East, and North Africa. Hundreds of millions of people in those regions would want affordable electric cars.

Scaringe and his team went to work designing a prototype. It took two full years. Then one night, as Scaringe describes it, as he lay awake in bed, he came to the terrifying conclusion that he was on the wrong track. "The middle classes in developing economies," he says, "don't go for leading-edge technology." The key for them is price. Whether it's cell phones or cars, they typically pile into commodity markets, which are often cutthroat. Scaringe might find himself in punishing price wars, with start-ups from Guangzhou and Bangkok to Islamabad.

He had to move above that fray and target wealthier customers. He considered the people who were buying luxury Teslas. They had money, often a lot of it. What's more, companies selling premium products to the monied classes can capture a fatter profit margin. And they aren't pushed into punishing mass production.

If Scaringe could sell premium electric vehicles at prices above $50,000, he'd have resources to load more technology into them—more sensors and autonomy, more tightly calibrated power management, and, as a result, a bigger range, perhaps topping three hundred miles on a charge. If Rivian could put together such a vehicle—a quality leader—the company could establish itself as a global brand.

That would be his pitch for the nerve-racking meeting with his chief investor, Mohammed Latif Jameel. He would explain that Rivian needed to shift its strategy, radically, and abandon two years of development—and then he'd need to ask for more money to start again. "I was sweating bullets," Scaringe says.

Mohammed, a savvy risk-taker, decided to double down—and

Scaringe got the funds. He promptly set out on a four-year project to design the electric car of the future.

IN THE EARLY years of this century, executives at Volkswagen faced a challenge. Toyota's new hybrid, the Prius, was blooming into an eco-sensation. With their electric engines engaged, Priuses purred along city streets, registering stupendous fuel economy. Many automakers, including VW, had long pooh-poohed the hybrid concept. Under the hood, the hybrid was an awkward mash-up of two engines, one powered by gas, the other by batteries (that filled up much of the storage space). A hybrid was like a fish with feet. Nevertheless, the Prius caught on, and it was capturing a growing green market.

VW, which was jousting with Toyota for world leadership in auto sales, had no electric answer to match the Prius. So the German giant came up with a counterintuitive response: diesel, the stinking, smoking engine technology that belched black exhaust from city buses. Marketing diesel as a green technology was akin to hawking cheeseburgers in a hospital's cardiology wing. Diesel engines, while more efficient than gas-powered ones, produce toxic fumes of nitrogen oxide and clouds of poisonous particulates. They contribute, far more than their share, to smog.

But that was the *old* diesel, VW executives said. Their new cars, launching in 2009, would run on a newer eco-friendly version: *clean diesel*. They promised loads of power and performance—much more than the Prius—and with far less pollution than traditional diesel engines.

VW's clean diesel initiative led to disaster, one that tarred the company's brand and swept out much of its leadership, including its CEO, Martin Winterkorn. Instead of creating clean diesel, VW cheated. It created software with a "defeat device" that activated pollution controls only when a car's engine was being tested. Consequently, the cars acted green and punchless in the laboratories, and then ratcheted up both performance *and pollution* on roads and highways. This was exposed in 2015, plummeting the entire company into crisis. Fines and other punishments, according to VW, eventually cost the company $33 billion.

But what's relevant here, for both the future of electric-powered cars and RJ Scaringe's start-up, is that the crime led VW, albeit belatedly, toward electric power. As penance, and part of the company's settlement with US regulators, VW agreed to invest $2 billion in electricity-charging infrastructure. According to Volkswagen, the goal of the initiative, Electrify America, is to build a "consumer-friendly charging network—to drive EV adoption by reducing charging anxiety." In the end, it was VW's diesel catastrophe that nudged the industry toward a future dominated by electricity.

The assumption throughout the auto industry is that electric cars are inevitable. China, the world's biggest market, is pushing for an electric fleet. Every major manufacturer is readying for the shift. Ford, for one, has entirely abandoned sedans fueled by gasoline (while clinging, for as long as possible, to its profit engines: gas-powered trucks and SUVs). Forecasts by Bloomberg New Energy Finance predict that electric cars will still be a niche market in 2022, selling 4.6 million units around the world, or about 6 percent of the total. But by 2030, the same analysts predict that number will reach 30 million—with nearly 40 percent of the sales in China.

The timing of this type of long-range forecast is notoriously unreliable. Investors who misjudge it will get burned. But in the long, or even the medium term, the timetable doesn't make much difference to the rest of us. Within a decade or two, electric and gasoline-powered engines are poised to switch roles, with electric dominating and internal combustion devolving into a niche technology. Gas-powered cars will likely hang on in the country, and among aficionados, much like horses a century ago.

Until very recently, maintaining an electric car was a pain—largely a labor of love. Electric cars cost more and, by many measures, performed worse. They had limited range. Finding charging stations was iffy. Their buyers paid a premium and put up with the downsides, largely for the allure of the new, along with the environmental benefit: zero tailpipe emissions.

Now, though, electric cars are rapidly catching up to traditional cars in price and performance, and the gains don't show any signs of slowing down. It's true that advances in chemistry-based technologies are harder and slower than the exponential leaps of digital systems. Nonetheless, research into batteries is booming, because the lightest and most efficient batteries will lay claim to rich chunks of mammoth mobility industries, from cars to phones.

This is bound to stir up troubles of its own, from the environment to geopolitics. Battery disposal is a mess, and mining for the minerals that go into them is rife with risks. Indeed, like oil and coal, the battery industry, at its root, is largely extractive. It feeds on rare earth metals such as lithium and cobalt. They're mined all too often by child laborers, some of them working twelve-hour days for less than ten dollars, according to Amnesty International. Cobalt, a bluish metal, boosts stability and storage capacity in batteries, both of which are crucial for electric cars. Half of the world's cobalt comes from the Democratic Republic of Congo, where child labor is rampant. In the coming electric car economy, regions like Central Africa could become as pivotal, and as explosive, as the oil-rich Persian Gulf today.

These issues have convinced a small and vocal contingent that other new fuels, such as hydrogen, should power mobility in the coming century. Hydrogen has great allure. It's the most abundant element in the universe, making up about 75 percent of matter. The only emission from vehicles running on hydrogen-powered fuel cells is water.

Hydrogen, though, has its issues. To achieve a competitive range of three or four hundred miles, a vehicle must store hydrogen at perilously high pressure. What's more, the industrial process to create the fuel feeds on large quantities of natural gas, which creates clouds of earth-warming carbon dioxide. If governments and global industry pushed for a hydrogen economy, no doubt many of these issues could be resolved. But most have already made their choice—and they're lining up behind the battery technology we already know, the kind we carry around in our gadgets every day.

THE NEWEST ELECTRICAL gadgets weigh several tons and roll on four wheels. They're among the star attractions in Los Angeles, at the LA Auto Show. Volkswagen, long past its diesel debacle, is diving into the arena. Its Audi unit is exhibiting its electric SUV, the $75,000 e-tron, which will later take a star turn in Marvel Studios' *Avengers: Endgame*. A VW concept van, Buzz Cargo, features a solar panel on its sunroof that could add as much as nine miles of range a day. And the company has teamed up with Amazon to sell and install home charging systems to e-tron buyers, at a cost of $1,000. Other electric models at the show include a couple of Kias, including an SUV, and the iNEXT, BMW's bid for the Tesla market. Its front grille is packed with sensors, enabling autonomous features, and it relies on cameras in the place of side mirrors—they're more aerodynamic.

But perhaps the strangest looking of the bunch is a silver pickup truck with a short bed. It has an emptier face than most trucks, with no grille in the front, just a white stripe running between the headlights. It looks like something Arnold Schwarzenegger's Terminator would drive. It's called the R1T. Standing next to it, wearing a broad smile, is RJ Scaringe.

This is the niche Rivian settled on: large, pricy vehicles for adventure and exploration. As he sees it, the premium market can be divided into two fields. One is luxury, the other more utilitarian. The electric luxury market, he says, is already getting crowded, with Tesla an established brand and Chinese competitors, led by Nio, targeting the same buyers.

But the other premium market for electric cars—the adventure market—remains largely uninhabited. There's a reason for that. Buyers tend to associate the sector with power, and power with the guzzling of gas. While most buyers use their adventure vehicles to pick up groceries or drop kids at school, they at least dream of adventure when buying one, whether it's fording fishing creeks in the Ozarks or plowing through a foot of snow above Lake Como. For this they want power. And like it or not, most people associate electric cars with "green," and "green" with "energy efficient"—which historically has been synonymous with "weak."

Scaringe's model for his adventure niche is the pricy outdoors retailer Patagonia. Founded by a fly fisherman, it retains a degree of authenticity in the adventure market. The tents it sells, the hiking boots and raincoats—they cost a lot and are supposed to last forever. That's the kind of brand Scaringe wants.

The design theme, Scaringe tells reporters at the LA Auto Show, is carefree, what he calls "invitational." In Rivian, as he sees it, the owner can do no wrong. The materials are tough and resistant, expressing a rugged quality to buyers, a bit like a $400 backpack or a pair of doeskin boots at Patagonia. "The test for me," says Scaringe, who has two small children: "Can you take the kids to the beach in it and not worry?"

In these early years, the market is still defined largely by the twentieth-century model, in which individuals buy, own, and drive the cars. Yet by the end of the decade, Scaringe predicts, many more of the cars will be served up by the hour or the day by commercial fleets. Increasing numbers of cars, including Rivian's, will drive themselves. But to reach that future as a going concern, Rivian has to get through the coming decade as something much closer to the twentieth-century model: selling cars to people who still have to take the wheel. For this market, the crucial factor is performance, and much of that hinges on the power supply.

The batteries are the heart of an electric vehicle, accounting for about one-third of the car's weight and also its cost. In Rivians, a panel of batteries lies on a low platform between the front and rear wheels. This gives the car greater balance and a low center of gravity. Gas-powered cars, by contrast, hold the engine, and much of their weight, off center, usually in the front of the vehicle. It's inherently less stable. More important, though, is the power in the Rivian's bed-size battery. It packs the electrical equivalent of a staggering 800 horsepower. That's more than *twice* the power of Ford's iconic F-150.

It's also much easier in electric trucks to harness the power where it's needed. Rivian's design features small motors powering each of the wheels. This enhances both handling on rough terrain and speed. Most customers, though, will focus on the biggest payoff from the battery: a range that will take them upwards of four hundred miles, and in the not-too-distant future—though Scaringe won't commit to a year—closer to five hundred.

The shape of that market, and the timing of it, from city to city, is still very much a mystery. The challenge for every car company, and especially tiny ones like Rivian, is to survive what's sure to be a turbulent transition to the networked, electric, and multimodal economy that awaits. This is where Rivian's so-called skateboard comes in. The innards of the electric vehicle—its battery, motors, and transmission—are deployed on a flat black structure the shape of a queen-size box spring. This is the business end of every Rivian vehicle, whether trucks or SUVs. Scaringe hopes to sell the skateboards as plug-in parts to a host of manufacturers. "They could be used for fleets of delivery vans, buses, you name it," he says. Even entrepreneurs, perhaps using custom manufacturing setups like Divergent 3D's, could assemble batches of vehicles for niche markets atop the Rivian skateboard.

The comparisons, for Scaringe and practically everyone else in new mobility, invariably return to earlier eras of information technology. In this case, Scaringe cites the rise of personal computers. In the 1980s, most of the computer industry, outside of Apple's small and fervent circle, took root on a common foundation: a standard line of semiconductors, most of them made by Intel Corp. The different companies, from Dell and Compaq to Sony, could make their own computer models. But they didn't have to bother making the computer's brain, the chips. Those were standard.

In the view of Rivian, and many others in the industry, the market for the next generation of electric vehicles will breed a new ecosystem, one with different specialists developing the various components—from software to touch screen windows. In this nascent market, Rivian is hoping to provide the engine and the power. That way, even if the $80,000 electric pickups don't jump off the lots, Rivian could still claw a path to survival.

Scaringe's strategy is finding new believers. In April of 2019, Ford plunked down $500 million for a piece of Rivian. The two companies will together develop an electric car for Ford—most likely powered by a Rivian skateboard. Earlier in the year, following the LA Auto Show, Amazon led a new $700 million round of financing, making it a leading investor in the start-up. Perhaps Rivian's skateboard will find its way into the coming fleets of delivery vehicles. In the do-or-die struggle ahead for auto manufacturers, that would be a big win.

4

Jurassic Detroit

In the summer between the two years at the Yale School of Management, students often take a corporate internship. Chris Thomas had an unusual destination in mind, and it led to a turning point in his life—one of many. He tells the story as we stand by the window of his expansive office on Woodward Avenue, gazing at the glimmering midday skyline of downtown Detroit. Thomas is barely forty, but he looks younger. He wears his hair combed back, which gives him a 1950s look, like a youthful Edsel Ford.

The people at Yale's internship office, he says, asked him where he wanted to go. He answered Detroit.

This was 2008. The auto industry was descending into a deep funk, perhaps a terminal one. The people at Yale wondered if he might not prefer London, or maybe Singapore. But he stuck with Detroit.

It was home. Thomas grew up just north of Detroit, in Waterford Township. The first in his family to go to college, he had studied at Michigan State. Then he'd gone off to work in banking in San Francisco, and from there to the war in Iraq. All this time, Detroit called to him. Though he may not have said it to the people at Yale, because it might have sounded like bragging, he wanted to contribute to his city's renaissance. "I love Detroit," he says. "I wear it on my sleeve all day long."

In the middle of the twentieth century this slice of Michigan was a land of fabulous wealth. Massive auto plants dominated entire neighborhoods in Dearborn, Pontiac, Ypsilanti. Detroit was one of the few places on earth where a generation of unionized factory workers could buy houses with two-car garages, even motorboats that they'd tow to the lakes. They could send their kids to excellent land grant universities, in nearby Ann Arbor, or in East Lansing, where Chris Thomas studied.

Thomas was born toward the tail end of the auto age, and he had known Detroit only in a state of decline and recurring crisis, some of it life-threatening. Five years before he was born, the Arab oil embargo of 1973 drew the curtain on Detroit's blockbuster era. As Americans hunted for fuel-efficient cars, the Japanese stormed into the market. They had a new industrial process, one grounded in data. "Kaizen," they called it—continuous improvement. They'd learned much of this science from an American industrial guru, W. Edwards Deming. No one in America had paid him much attention.

The Japanese invasion was Detroit's rude welcome to global competition. In the following decades, while Thomas was growing up, companies migrated work to the nonunion South and to Mexico. Entire supply chains jumped to China. Detroit hollowed out. Months after Thomas applied for his internship, GM and Chrysler almost died in the financial crash of 2008.

Thomas's mission, from long before his days at Yale, has long been to help rescue Detroit. The city can never hope to dominate as it once did. Plenty of other hot spots are growing, in Stuttgart, Tokyo, and LA, and on the software side, in Silicon Valley, Tel Aviv, and Shenzhen. But he insists that Detroit, with the largest transportation cluster in the world, should have a major role. "It's not a birthright," he says. "It's something we're going to have to run at very hard."

Thomas is a cofounder of a venture investment firm, Fontinalis. It focuses exclusively on the next stage of mobility, and has invested more than $200 million in many of the technologies we describe in this book, from AI to dockless bikes. Fontinalis was the first venture firm to follow this sector, though now it has loads of company. One of his cofounders is William Clay Ford Jr., the chairman of Ford Motor Company and a great-grandson of its founder, Henry Ford. In this sense, money from the first age of manufacturing is seeding the second.

Thomas met Ford the summer of his internship. That's the story he tells as we look out the window. It was a grim time at the company, a season of layoffs and dark premonitions. Auto sales were slowing down as the once booming real estate market turned from white hot to glacial. Months later, the collapse of Lehman Brothers would trigger the Great Recession and plunge Detroit's automakers into existential crisis.

On his first day at Ford, Thomas was dispatched to the company treasury, a logical assignment for the former banker. He soon found himself sitting at a cubicle and facing what he calls the "most boring job you can imagine." It had to do with inventory management. He hated it.

So he wrote emails, sending one to every top executive at the company. He asked for a half hour of their time, to hear about their jobs, their view of the company, and to learn, if not the answers, at least the questions that needed to be asked. A few of them invited him upstairs. They had friendly but inconclusive chats.

Thomas was sitting in his cube a few days later, grappling

with inventory, when he got a call. It was Bill Ford, as he was known, the executive chairman. Ford asked Thomas to hop up to his office. He had a half hour.

The chairman greeted the intern with a broad smile. "Tell me how much you love your internship," he said.

Thomas responded that he hated it.

Ford took in that uncomfortable news. "Well," he said, "we have twenty-nine minutes to talk."

So they talked. Thomas told him a bit of his story, his upbringing in Detroit, and studies in Michigan State, how he'd then gotten a job as an investment banker in San Francisco, just as the dot-com boom was crashing. Then came the terrorist attacks in 2001, and the march up to the war in Iraq. After his younger brother, Scott, called him to say he was enlisting, Thomas followed suit. He'd walked right down to the recruiting office in San Francisco still wearing his banker suit, and a year later was running communications at a post north of Baghdad. Then came graduate school at Yale. He told the chairman that he was eager, even desperate, to find more interesting work at Ford. He wanted something with greater impact than shuttling inventory hither and yon.

Ford listened. When the half hour was up and they shook hands, Thomas held on to the chairman's hand as he made a request. He asked to work on the most interesting project in the entire company. Ford was noncommittal and wished him a good weekend.

The following week, Thomas was reassigned. He would be doing a job, the chairman told him, that most people didn't know existed. He was to join a group at Ford that was mapping out the future of mobility in the world's megacities. It was an informal team. It had no budget. "Skunkworks," Ford called it—the moniker for stealthy innovation units, named after the small teams

Lockheed Martin assembled during World War II. The team members all had other jobs, and they carved out time to brainstorm about next-generation mobility in places like São Paulo, Shanghai, and New York City. Only months before the global financial crash, this informal group was busy sniffing out the future.

This team envisioned the impact of smart, networked, and increasingly robotic transportation. These new trends and technologies, they saw, would give birth to an entire ecosystem of mobility, including fleet operators, battery companies, mapping apps, an entire universe of software, and a new menagerie of exotic conveyances with wheels and wings.

At summer's end, Thomas returned to New Haven, his head swimming with all the possibilities in mobility. He and his classmate Chris Cheever were intent on creating a start-up for this new industry. They considered engines and cars and batteries. But they couldn't figure out which opportunity to pursue.

Even if they settled on a strategy, they wondered, where would the financing come from? The world's economy was dive-bombing, and what money there was trickled into social networks and apps for the iPhone, which at that point was barely a year old. Entrepreneurs could launch businesses on their laptops. Who would ever spend money on an industrial start-up, and in Detroit, of all places?

That very question pointed to an unlikely business opportunity. The following spring, at the nadir of the financial crash, Thomas again knocked on Bill Ford's door and made a pitch.

Detroit, he said, had been a global capital for the age of the automobile. It should lead in the next century as well. To that end, industry leaders, including Ford, should invest in the next generation of mobility start-ups. Such a venture might have sounded pie-in-the-sky a few years earlier, when Detroit's cars and trucks were jumping off the lots. But in the spring of 2009, with both GM and Chrysler in bankruptcy, and Ford limping badly, the future suddenly seemed close, perilously so. Investments in new mobility held out a possible lifeline. They might even be lucrative.

Ford provided the initial capital for Thomas and Cheever's venture company, Fontinalis. The two men proceeded to set it up in an office in downtown Detroit. For the next few years, Fontinalis had the venture market for mobility almost to itself.

In the first years following the crash, Detroit looked downright dystopian. It contained block after block of derelict housing, row houses with insulation bleeding from holes in aluminum siding, doors knocked down. People were suffering. The most prominent pedestrians on downtown streets, it sometimes seemed, were homeless people pushing their tattered belongings in repurposed shopping carts.

The city had declared bankruptcy, and many wondered if the treasures from its famed Institute of Arts, perhaps even Diego Rivera's Depression-era murals of Ford's River Rouge plant, would be auctioned off to pay creditors. Detroit was so empty and poor that the city struggled to deliver basic services, like plowing snow and operating school buses. The mayor pushed, unsuccessfully, to tear down the outer neighborhoods and shrink Detroit's footprint. His idea was that everyone left could huddle closer together.

In the universe of mobility hot spots, postcrash Detroit was a forlorn wannabe. To find promising start-ups, Thomas and his partners at Fontinalis had to travel to places like California and Massachusetts. Yet nearly everywhere they went, they came across exiled Michiganders. It stood to reason. People from the Motor City were more likely than the rest of us to be smitten by cars and the industries of mobility. However, the opportunities—the investment and talent—were elsewhere. So they left.

One of those they met was Karl Iagnemma. Born in Detroit

and educated at the University of Michigan, Iagnemma had gone east, to the Massachusetts Institute of Technology, for his PhD in mechanical engineering. A polymath, he had published short stories and a novel before launching a start-up in Cambridge to develop software for autonomous cars. Fontinalis was an early investor in the company, NuTonomy. Years later, in 2017, the venture firm raked in a fat return when Delphi, the auto parts giant, swooped in to buy it for a reported $450 million.

There were plenty of other successes, too, most of them far from Detroit. ParkMobile, an Atlanta smart-parking start-up, sold to BMW, and SPLT, a New York carpooling platform, sold to Bosch, both for undisclosed prices. Verizon paid more than $900 million for Telogis, a California company whose software manages connected cars. "I never expected the mobility market to become this big this fast," Thomas says.

Fontinalis enjoys a privileged position in the booming market. However, most of the jobs its portfolio companies create, whether in cybersecurity or fleet management, go to a privileged subset of talented engineers and programmers. That's the nature of tech revolutions. Fewer people get a lot richer—and most of them live in tech havens hundreds of miles from Detroit.

For industrial cities to make an enduring comeback—which was Chris Thomas's driving goal from the get-go—manufacturers have to succeed. True, big automakers will never relive the Motor City's heyday, when tens of thousands of well-paid workers streamed in for their shifts. The auto plants are leaner now, and equipped with robots. But still, auto manufacturers hire workers by the hundreds, not by the ones and twos. Do they have a future in cities like Detroit?

SUMMER OF 2018. Downtown Detroit is back, and it's infinitely livelier than a decade earlier. Local business leaders, including the founders of Quicken Loans and Domino's Pizza, have bought entire blocks of real estate and are refurbishing them. The streets are teeming with young professionals. The restaurants are packed. People zip around on Bird scooters.

On a bright June morning, Ford's chairman, Bill Ford, invites the press to a once grand railroad station in Detroit's Corktown neighborhood, just west of downtown. Michigan Central Station, empty since the last train pulled out in 1988, is covered with graffiti, its windows shattered. Now, with funding help from the state of Michigan, Ford announces plans to turn the station into a downtown headquarters for development in a host of mobility technologies. Teams of scientists and engineers will design autonomous cars here, he says. In-house entrepreneurs will launch ride-hailing services and new delivery businesses. Ford vows to create in Detroit the "mobility corridor of the next 50 years." It's as if he's channeling Chris Thomas.

Legacy automakers, including Detroit's Big Three, have been girding for this technological shift. In fact, new mobility has become something of a legacy mantra. All of them, from Germany's VW to Great Wall in China, are working on autonomous technologies. They have lined up partners; invested in delivery services, scooters, and sensors; and poured money into software start-ups from Silicon Valley to Tel Aviv.

Ford, to name one, has earmarked investments of $11 billion to position itself for new mobility. It spent $1 billion for a share of Argo AI, a Pittsburgh autonomous car company. Ford also owns a Silicon Valley company called Autonomic, which builds cloud computing infrastructure for transportation, and TransLoc, of North Carolina, whose software manages urban transit. The company has a partnership with Lyft to develop autonomous cars. It even launched a high-profile pilot program in

Miami to deliver Domino's pizzas in autonomous cars. And Ford is building an electric car with RJ Scaringe's Rivian.

Change is inevitable, the automakers agree. The statements they issue on the matter have become familiar to the point of cliché. On an earnings call with analysts in 2018, Ford's CEO, Jim Hackett, said, "We see ourselves not just as a provider of mobile solutions, but also as an orchestrator of digital connections." Other auto executives mouth similar sentiments in much the same language.

The question isn't whether legacy automakers intend to embark on this difficult transition, but if they're able to. RJ Scaringe, scrawling on his whiteboard at Rivian headquarters, about an hour west of Chris Thomas's office, summed up the simple and grim equation that mass manufacturers face: "If through the efficiencies of networking and sharing we can double the usage of our vehicles, we'll probably need fewer of them."

As we saw at Kevin Czinger's Divergent 3D factory in California, new technologies will enable entrepreneurs to set up minifactories and produce batches of vehicles for niche markets. Those, too, threaten to erode the market for mass-produced cars and trucks and autonomous pods.

But perhaps the biggest challenge these legacy automakers face is a shift in the very nature of their business. Consumers, increasingly, will be buying more services and fewer machines. The market will be one for miles. The auto business of the future will involve establishing relations with millions of customers, feeding on their data, anticipating where they want to go, and getting them there, perhaps providing a cappuccino, a virtual reality escapade, or a foot massage en route.

For this, companies like Ford, GM, and Toyota are planning not just to manufacture vehicles, but to operate fleets of them in cities around the world. Those are some of the "mobile solutions" and "digital connections" Hackett talks about. It's why auto executives are consuming software companies and fleet-management firms with a survivalist fervor. "There's a tremendous amount of fear," Thomas says.

But what advantage does a big car company have over a pure service outfit, like Uber, Lyft, or China's DiDi?

Let's stick with Ford. In a good year, like 2017, its 200,000 employees build nearly seven million cars and trucks. Revenue tops $150 billion, with about 6 percent of that, or $9 billion, jingling down as profits. That size gives a behemoth like Ford all kinds of advantages over start-ups like Rivian, including global distribution and immense leverage over suppliers. When it comes to making and selling things, legacy automakers are well positioned.

Size, however, can be a problem. While traditional automakers run the biggest industrial operations in history, their profits hinge on scale. When they operate below capacity, their sales fall faster than their costs, and those billions in profits turn quickly into losses. Black turns red, and the numbers, like everything else in the auto business, grow enormous. Chris Thomas was seeing the beginning of this destructive pattern during his few unhappy days of inventory management in the summer of 2008.

So when it comes to an auto giant changing its business model and turning to services, its own enormous scale turns into a ponderous load. How can the company maintain its revenue and support its core business while it's hurrying to build and launch scores of start-ups to disrupt it?

THIS HAPPENS IN every industrial revolution. Burly incumbents, as Clayton Christensen writes in *The Innovator's Dilemma*, have to figure out how to survive while disrupting their own businesses. It's a recurring theme, especially in technology. Perhaps the best example for what the auto giants face is the story of IBM, the onetime titan of computing.

In the middle of the twentieth century, IBM ruled. In its industry, you could argue, it was more powerful than Detroit's Big Three, Toyota, and Daimler combined in the global auto market. IBM defined computing.

In its glory days, IBM had a lot in common with today's car companies. It manufactured big machines. It also ran an unrivaled distribution network—legions of salespeople, mostly men, wearing the trademark blue suits. They visited corporate customers and sold them machines. That was the core of the business. Servicing the machines and the software that ran them? That was included in maintenance contracts.

This was a world of immense waste, and IBM benefitted hugely from it. Each company had to have its own computers, much the way each family in an urban sprawl needed its own cars. There was no way to share them. What's more, each company required enough computing to handle the busiest months of the year, like tax season, or for retailers, the holiday rush. So, much of the hardware they bought spent the slow seasons nearly idle in its refrigerated rooms. They were like cars killing time in driveways and parking lots.

IBM's machines were expensive. But computing itself was getting exponentially cheaper. Moore's law, conceived in the 1960s, predicted accurately that computing power would double and its cost would fall by half every eighteen months. For IBM, this was unsustainable. The company had huge factories pumping out big computers, and highly paid sales teams. These were fixed costs. Yet in the 1980s as personal computers spread, customers were seeing that they could get a $2,000 PC to do much of the same work as IBM's more expensive machines.

By the early 1990s, IBM was near collapse. The board hired an outsider, Louis V. Gerstner Jr., to rescue Big Blue. Gerstner knew little about computers. He had worked as a top executive at American Express, and then been the CEO of RJR Nabisco, a food conglomerate. But he knew enough about markets to see that IBM needed to find new ways to make money.

His answer, much like that of the auto companies today, was to turn the company toward services. If computing power was becoming a commodity, IBM would make its money helping customers make smart use of it. Over the following decade, IBM turned away from manufacturing and focused on services and software. It was a wrenching transition. IBM shrank and laid off tens of thousands of employees. Revenue fell. Still, Gerstner and his team pulled it off.

IBM survived, but no longer dominates. During its transition, it found itself facing a host of companies that were native to the new era of computing. Software companies like Microsoft didn't make machines, and they were unburdened by those immense legacy costs. Google, native to the Internet, didn't make anything at all.

Google's parent company, Alphabet, at this writing, is worth seven times as much as IBM. The key—and this point is especially relevant to mobility—is that while IBM no longer sells loads of machines, its core business is still tied to technology, which keeps getting cheaper. That punishes sales. For Google, by contrast, technology is merely a tool. Unlike IBM, Google benefits as prices drop, and it uses this ever cheaper and more powerful tool to sell something else entirely: advertising. Facebook works on the same dynamic.

Now consider the mobility revolution. We don't know which companies will emerge as champions of connected cars, nor which automated technology will come out on top. We don't know at this point what these rolling units will look like, nor what they'll be called. But one thing we do know, with near certainty, is that miles will get *cheaper.*

That's what digital technology does when it infests an industry. It takes something that was scarce and makes it both plentiful and cheap. The exponential growth of computing power feeds this dynamic, as does the very nature of software and silicon. Once software takes over a job, it can be replicated infinitely at negligible cost—and devour entire industries.

How to survive in an industry of crashing prices? One approach is to build a business that sits on top of all the risk and hard work that other companies do, and sells a service. This is what Google and Facebook pulled off in computing.

It's also what Uber and Lyft are busy building in mobility. They don't buy steel and glass, they don't haggle with the United Auto Workers. They don't have to compete against Daimler or VW or a host of new manufacturers from China. At the same time, they don't have to lay down asphalt or build bridges. Governments handle that. So the ride-share groups are divorced from the costs, risks, and responsibilities of making stuff. In the wild, they'd be considered a particularly insidious variety of parasite.

The ride-share companies are simply apps on phones. (The ultimate cheapskates, they don't even invest in mapping technology, outsourcing much of that work to Google.) Yet unlike the auto companies, they forge intimate contact with millions of customers. They mine rich streams of customer data and, more important, they build tens of millions of billing relationships. In any economy, new or old, a beeline into a customer's bank account is pure gold.

WHILE EASING OUT of manufacturing, car companies will be competing for different markets and needs. They'll have to innovate new types of vehicles.

J Mays—"J" a name, not an initial—was Ford's lead designer for eighteen years. He reintroduced the Mustang and shaped the popular F-150 truck. Now in his sixties, with short white hair and square-rimmed glasses, he trains his designing eye on dishwashers and refrigerators at Whirlpool.

But he still spends a lot of time imagining the shape of cars to come. The design challenge, he says, will be entirely different as we move from an ownership to a passenger economy. For most of its history, the automobile was an appendage of a person. Rolling in a car, perhaps an elbow out the window, the radio playing loud, a driver expressed power and status and personality. For its part, a car promised speed, freedom, and, to one degree or another, sex. Mays's Mustangs certainly aspired to that.

Jump ahead to, say, Paris or Seattle in a few years. A graduate student wants to cross town. She calls an autonomous fleet car. Most of the time, she's not investing in self-expression or status. Instead she wants efficiency, and preferably a pleasant experience. "It might feel like more of a cocoon, which enables us to float to our destination," Mays speculates.

The service, he says, should allow people to climb into the vehicle and experience "a little vacation." Maybe it will feel like they're lost—more a freedom of the mind than one of movement—before they "come out of the vehicle joyful, without the drudgery."

How do you design *that*?

Who designs it? Will it be the auto manufacturer, outfitting chambers with comfy pillows and satin (spill-proof) surfaces, fabulous Sensurround audio and digital screens that look realer than life? Maybe. But another service might take the traveler a step further, and into virtual reality, perhaps to a tropical beach, or Disney World. This would constitute yet another case of software hijacking a business from the physical world.

Software and silicon will continue to battle for a greater share of the car, from entertainment to navigation. Asutosh Padhi, a McKinsey & Company analyst, predicts that the amount of software content in the car will triple in the next fifteen years.

Piece by piece, software firms are out to conquer the car. A Boston-based company called ClearMotion, for example, is focusing AI on automobile suspension; the idea is to anticipate the bumps on a highway before they come, and then to respond appropriately. The company compares this suspension AI to noise canceling in headphones. Investors like the idea and have poured in $270 million in venture funding. If it works, it lifts yet another piece of the car budget from engineers in places like Detroit and delivers it to computer scientists on one of the coasts.

As cars turn into rolling assemblies of computers and become increasingly autonomous, they will in effect transform into a new and different species. As this happens, the numbers associated with them, from price to life expectancy, will change dramatically. This will make it even harder for automakers to plan and budget.

The next generation of cars—if that's what we'll call these electrical and semiautonomous units—might cost $150,000 each when they first appear. But they could cost only a third as much five years later. At the same time, the cars will be electric, and they may have one-tenth as many moving parts. So they'll be far more reliable, and they might travel for millions of miles. The fleet owners, having paid richly for them, will likely keep investing in them, buying new parts.

It's a different business, in some ways more like today's buses. Most of us don't dream of owning one. We leave that to companies or municipal transit agencies. Further, most people have no idea how much they cost. It's not relevant to them. Unlike cars, buses run for decades, because fixing them is a whole lot cheaper than buying new ones.

In this sense, buses provide a glimpse into the auto markets of the future. That future features serious jousting involving the Chinese.

SEPTEMBER 2003. RYAN POPPLE, an angular twenty-three-year-old Chicagoan, was working as a US Army tank commander in Iraq. His logistical challenge was to make sure that legions of fuel-devouring tanks always had enough to drink. This meant establishing supply chains to move tons of gasoline as efficiently as possible. It was a sensible job for Popple, who had majored in economics at William and Mary, in Virginia. But it was hard. A tank battalion powered by internal combustion is a ravenous beast.

Managing tanks got Popple thinking about the economics of mobility, and especially about fuel. While on his tour in Iraq, he could hardly escape the nasty byproduct of fuel, the diesel smoke that tanks belched into the desert air. It was toxic—literally sickening.

Popple left the army after five years. His goal then was to work for a cleaner, more sustainable planet, and to help wean humanity from what he viewed as its suicidal addiction to fossil fuels. He still regards this as a generational challenge. He was born in 1977, he says. His parents grew up at the postwar height of the American century, when the growing middle class in the United States could buy just about anything and blithely dismiss the notion of "sustainability" (assuming the word ever came up). His own generation, he says, may be closer in worldview to its Depression-era grandparents, who had learned to make do with scarce resources and frowned on waste.

Upon his return, Popple settled briefly in Cambridge and got his MBA at Harvard. From there he traveled west, to California, where an entrepreneur named Elon Musk was starting to build electric cars. This was before most people had heard of Tesla.

Popple ran finances for Musk's company as it developed its first electric roadster. Then he moved to the renowned Silicon Valley venture firm Kleiner Perkins, where he specialized in green investments.

One of Popple's venture investments was an electric bus start-up called Proterra. As a leading investor, he sat on the board. And when the board dismissed the founding CEO (which happens with regularity), Popple found himself entrusted with the job, on a temporary basis, while the board hunted for a permanent replacement.

He's still running the business.

It wasn't until Popple took control of Proterra, he says, that he realized the company had been aiming far too low. Its initial focus was on the "green" market, starting with environmentally conscious towns, places like Portland, Oregon, or Madison, Wisconsin, where a government might pay above the market rate for a few electric buses. The new buses would help clean up the city a bit, bolster its brand as forward thinking, and send the right signal to green voters. Proterra hoped to sell one thousand vehicles per year into these niche markets, enough to make it as a midsize company.

When Popple studied the numbers, he came to a much different conclusion, and a larger one. The trends were clear: Batteries were gaining in power, much faster than the industry's estimates. The price per kilowatt-hour was plummeting. Early in the 2020s, electric engines would match diesel in range and price. All other things equal, who wants filthy diesel if there's a cleaner option? He saw that electricity, in the coming decade, could power entire industries of large vehicles, including the global bus market. And Popple predicts it will. "I can't see anyone ordering a diesel after . . ." He runs the numbers in his mind for a second or two. "2025," he says.

Proterra's headquarters (and battery plant) is just down the road from the San Francisco airport, a ten-minute walk from the BART train. Popple often rides his bike to work. It hangs on the wall behind his desk. He speaks earnestly, with a strong analytical bent and a focus on the big picture: the survival of our species.

Much of humanity's future, he argues, hinges on how we manage our cities. They already house half of humanity, and that share is sure to keep growing. This is a good thing, because cities are greener than the suburbs or the country, mostly for one reason: people jammed together don't need to move nearly as much. The key for our future, he says, is to figure out how to operate our cities, to make them cleaner, safer, and healthier.

Such a cleanup can start with urban buses. Only a tiny minority of the world's cities run trains or subways. For everyone else, public transportation comes down to some sort of bus.

From Popple's former military perspective, urban buses are like several platoons of tanks circulating continuously on city streets. But instead of the empty desert, these behemoths are blowing diesel smoke into the faces and lungs of millions of closely packed human beings, including children. Billions of particulates known as PM 2.5 float in the exhaust from the diesel engine, Popple notes. These are toasted microparticles, each one smaller than 2.5 microns. About forty of them would fit into the period at the end of this sentence. "The more we look at it," Popple says, "the scarier it is, especially for children." These microparticles are catalyzing agents for cancer. A study of demented dogs in Mexico City suggests that such particles may work their way into the brain and wreak neurological havoc there.

Many cities are responding to this environmental challenge by replacing diesel buses as they retire with others powered by cleaner natural gas or batteries. At this writing, diesel still holds an advantage in price, but it's shrinking. A diesel bus costs about $400,000; one powered by natural gas, $500,000; and one of the electric Proterras coming out of its South Carolina plant, $600,000. But city officials also have to consider the performance of the vehicle, along with its life expectancy and annual maintenance costs.

This is where the electric industry faces its biggest challenges. Los Angeles provides a case study of how a big bet on alternative fuels can go awry. The city has long been eager to reduce smog, and it has pledged to reduce emissions from public vehicles to zero by 2030. So when a delegation from a Chinese industrial giant called BYD (the acronym, the company says, for "Build Your Dreams") arrived in Los Angeles with a plan to manufacture electric buses in nearby Lancaster, they got a warm welcome. After all, BYD was promising manufacturing jobs, perhaps a thousand of them, in a cutting-edge green industry. This was catnip for politicians. And the company seemed legit. Warren Buffett, the Oracle of Omaha, was a leading investor.

BYD didn't have a history making buses. Launched in Shenzhen in 1995, the company started by making batteries for cell phones. Later it moved into electric cars and, eventually, buses. Its timing appeared spot-on. As smog thickened around Chinese cities, the government was vowing to electrify transportation. This promised a near-limitless market for electric buses. (In the single year of 2017, just for context, Chinese cities purchased 87,000 e-buses, according to McKinsey & Company. The entire American fleet is only 70,000, with fewer than 6,000 in its largest market, New York City.)

In September 2008, just as global markets were crashing, Buffett's Berkshire Hathaway holding company paid $232 million for a 10 percent stake in BYD. In the company's billionaire founder,

Wang Chuanfu, Buffett saw a leader in the mobility revolution. His longtime business partner Charlie Munger, he said, described Chuanfu as a combination of Thomas Edison and Bill Gates.

Perhaps Wang Chuanfu will scale those heights—but not without encountering bumps along the way. In the decade following Buffett's investment, BYD set up its factory in Lancaster, and municipalities in Southern California, including LA, ordered $330 million in electric buses, trucks, and industrial equipment from the company. According to a 2018 *Los Angeles Times* investigation, the buses brought little but troubles. Their charges were too weak. Buses faltered on hills. The range, listed at 155 miles, was in actuality far less. Drivers took them to be recharged, on average, after only 58 miles. They also broke down far more often than the diesel stalwarts. A BYD vice president blamed the troubles on LA Metro, for making too many stops and driving up hills that were too steep.

You might think that for Ryan Popple, quality concerns at his biggest competitor would be welcome news. But each faulty electric bus, no matter who makes it, tars the reputation of the upstart industry. The same dynamic threatens other new mobility industries. If an autonomous car turns against traffic on Broadway, or if an airpod tumbles into the sea, people will question the viability of the technology.

To dispel the doubts surrounding battery-powered buses, Popple runs them through tests. In one trial in 2017, a Proterra bus smashed a record by running 1,100 miles on a single charge. This was hardly in normal conditions. It had no passengers or stops, and it ran on a flat track in Indiana. Still, it supported Proterra's promise that its buses, on a single charge, could handle 250 miles on stop-and-start urban routes, while loaded down with scores of passengers. In the following months, the company landed sizable contracts for Seattle and Chicago.

But Popple, an engineer, gets especially excited imagining the school bus, or legions of them, in the energy markets. Nearly half a million school buses operate in the United States. That's more than seven times the size of America's combined urban fleets. But school buses operate at full capacity during only two periods of the day, an hour or two in the morning, Monday through Friday, and again in the afternoon. The rest of the time, most of the buses are sitting around idle.

This is where the energy markets come in. If you focus on the batteries of these buses, each one is an electricity asset—a powerful storage unit. An electric bus soaks up the cheap midday volts pouring from solar panels. Later in the evening, when people are back home and turning on their TVs and air conditioners, it could deliver some of the energy back to the grid. Coordinating the supply of alternative energy with the market demand is an enormous challenge for the industry. Electric school buses could help.

What's more, these storage units are on wheels. They can go where electricity is needed. This could prove helpful in disasters. "Think about when Hurricane Maria hit Puerto Rico, when no one had lights and the hospitals had no power," Popple says. School buses with megabatteries "could have made a big difference."

THIS IS WHAT happens as mechanical vehicles transform, step-by-step, into electronic appliances. The old barriers among businesses fade away. A school bus rolls into the electricity business. The hospitality industry moves on to taxis. It's a giant reorganization, a free-for-all. It creates opportunities for newcomers—and risks for big incumbents, whether utilities or car companies.

This brings us back to Chris Thomas. He no longer works in the downtown office high above Woodward Avenue. In fall of

2018, he left Fontinalis to dedicate himself to a new start-up, Detroit Mobility Lab. "When you look at the global auto ecosystem," he says, "everything is here. It's a legacy we can leverage. But how do we win? It's a question I ask myself all the time." Then he answers his own question: "We need talent, company creation, and investment." So his mission, he says, is to build a local talent base to keep Detroit in the middle of the mobility economy. A Boston Consulting Group study estimates that the new mobility economy will create 100,000 new jobs in the United States. But many of the best jobs, about 30,000 of them, will be in fields like AI, robotics, and cybersecurity. Most of the talent in those fields, in the American market, resides on the coasts.

To nurture know-how in Detroit, Thomas's group plans to build a school—a physical one—called the Detroit Mobility Institute. It will educate the city's professionals and tradespeople in the twenty-first-century skills they'll need for mobility jobs. The executive director, Jessica Robinson, who came from Ford, says the institute, working with university partners and industry alike, will be offering a new degree, Master of Mobility, by 2021.

If successful, some of the institute's talent, no doubt, will feed the legacy automakers. And yes, a few of the graduates might launch start-ups that disrupt those auto centenarians, or even supplant them. That's how healthy ecosystems work.

5

Helsinki: Weaving Magic Carpet Apps

The town of Masku, nestled near Finland's west coast, lies on the main highway linking the city of Turku and the northern town of Oulu, near the Arctic Circle. To the east stretches a vast Baltic archipelago, with more than six thousand islands. To the west is a constellation of ponds, wrapped in dense forests of birch and oak. Masku, with its population of eight thousand, sprawls on both sides of the highway, occupying land as carelessly as small towns in Texas or Alaska. Space is abundant. People get around in cars.

This makes Masku a royal pain for a schoolgirl who has to get to soccer practice across town. The girl in this case, in the early years of this century, was a tall blond-haired athlete named Sonja Heikkilä. In the scarce months of decent weather, she'd bike, and when she was fifteen, she bought a motor scooter. That helped. But the scooter disappeared into the garage for the long winters. Moving around in Masku was a headache.

In 2008, by then eighteen, Heikkilä moved to the capital, Helsinki, to study engineering at Aalto University. Compared with Masku, Helsinki at first seemed like a public-transit nirvana. A steady flow of trams snaked around the lakes, through the massive squares, and to the port. Helsinki also operated legions of buses and even a metro line. But Heikkilä still struggled. She remembers standing on dark corners after soccer practice, squinting through sleet and snow to see which bus was coming. "The weather was lousy most of the time," she says. Finally, she was fed up to the point that she bought a beat-up Fiat Mondo—which proceeded to break down on a regular basis.

Sonja Heikkilä is a serious person, and brainy. A problem solver. Like billions of us around the world, she faced mobility challenges nearly every day. They were stealing her precious time and emptying her bank account. Even with all its trams and buses, she realized, Helsinki was still a car town. Those without reliable cars formed an underclass. "It's not fair that car owners have the freedom of mobility, and others don't," she says.

Heikkilä, of course, had a smartphone. At the touch of a finger, she could call up music, weather, and maps, and text friends. But the one app she needed the most—mobility—was missing. It didn't exist, because transit data in Helsinki, and just about everywhere else on earth, came in separate streams.

This isn't to say that the data was useless. Bus data, for example, fueled a valuable service. As early as the 1990s, Finns could call up a text service on the rudimentary cell phones of the time to see when the next bus was coming. If it was fifteen minutes off, they might stay out of the cold a bit longer and have another cup of coffee.

This was valuable data, but it was isolated. Taxis and trams had their own data. So did the highway authorities, who could track motor traffic on the seven highways streaming into the capital. At a control center in downtown Helsinki, much of the data came together, providing a minute-to-minute look at movement in the capital. But there was no app for the people who needed it most: those who needed to get from one place to another every day. They were traveling virtually blind.

Heikkilä imagined a single app that would not only let her see and measure various mobility alternatives, but also pay for them. Her app would work so smoothly that people would ditch their cars.

In many ways, the mobility app Sonja Heikkilä dreamed up fell into line with other trends in the digital economy, such as music. When she first arrived as an eighteen-year-old at the university, in 2008, she and most of her classmates owned their music. They downloaded songs from legal services, like Apple's iTunes, or from pirate sites like Kazaa. In the quaintest cases, they copied music from CDs. (In 2008, just to note how fast things have changed, many people still listened to these downloaded songs on iPods.) Within four years, however, the entire business changed. Songs were no longer something to own. More and more, people subscribed to music services, on apps like Spotify and Pandora. For about $100 a year, the price of eight or ten CDs, music lovers could waltz through life carrying a weightless jukebox that played practically any song they could dream of.

If streaming was "music as a service," Heikkilä's app would provide "mobility as a service." This concept had also been percolating elsewhere, but she developed a concrete proposal for Helsinki. She called it MaaS. Instead of demanding a song or a symphony, the user would request a time and a place. The app, running various streams of mobility data through analytics programs, would propose the optimal combinations of transport, each with its projected arrival time. The app would even take care of payment seamlessly, so that Heikkilä would not have to fumble for change when she hopped on a bus.

Helsinki is not a big town. With a population of 650,000, it's about the size of Nashville, Tennessee, or Portland, Oregon. Its metro population represents more than one of four Finns, but barely reaches 1.5 million. In this midsize market, Helsinki's transportation insiders, whether at the university or at the metro authority, all know one another. Little surprise, then, that Heikkilä's idea spread quickly among them. It wasn't long before the city's transit department jumped on it and commissioned the young engineering student to take her proposal further and create an in-depth feasibility study on MaaS. Over the next several years, Heikkilä extended and refined her vision, finally publishing it in 2014 as her master's thesis.

The heart of her concept was to provide a service to rival owning a car. It was clear, she wrote, that no single service could match the automobile by itself. No matter how much money the city poured into trams and subway extensions, mass transit could never reach every home. In the endless sunlight of Finnish summer, bikes were terrific. But they provided only a partial answer, and they were certainly not for everyone.

No, to compete with the car, the mobility app would have to offer *every* people-moving option, from cabs and the subway to dockless scooters. It would be multimodal. Once such an offering existed, Sonja Heikkilä allowed herself to dream, Helsinki would lead the rest of the world into smart mobility. Green spaces would expand. Her app could change the life and even the topography of the entire city.

The mobility app, of course, would be a lot more complicated than a music app like Spotify. Moving molecules takes a lot more work, and money, than shooting ones and zeros through digital networks. But as Heikkilä pointed out in her thesis, plenty of logistics companies, including DHL and FedEx, had advanced the mathematics of movement. They could determine, within a second or two, the best route to move a hair dryer from a warehouse in Taiwan to an apartment in Teaneck, New Jersey. The best path might involve six or seven different conveyances. In their own way, these giants already had mastered multimodal.

So had manufacturing giants. Companies like BMW and Samsung orchestrated the movements of giant supply chains of components that went into the cars and phones and microwave ovens they produced. This was another variation on the same theme. So there existed plenty of expertise in moving complex collections of molecules from point A to point B at optimal cost and speed. Now it was just a question of applying that science to the movement of human beings in Helsinki—and then building a service business around it, and baking it into an app.

As Heikkilä developed the plan, car-sharing services like Uber and Lyft were popping up in cities. Bike-shares, a novelty a decade earlier, were everywhere. With every new transit option, the MaaS concept grew more powerful. As Heikkilä saw it, service providers would sell different levels of subscriptions. Expensive ones might give customers more access to taxis and car-shares, and eventually, autonomous pods. Cheaper plans would be heavier on bikes, scooters, and trams. Users could toggle the controls, asking for the fastest routes, the cheapest, the greenest, even the most scenic.

When she published her thesis in the spring of 2014, she predicted that the movement of human beings could undergo a revolution by 2025. It would be cleaner, greener, faster, cheaper, and more fun. Heikkilä, of course, wasn't thinking about only Helsinki. She wanted to revamp human mobility in cities around the world.

MID-MAY IN HELSINKI. Blustery winds blow in from the Baltic Sea, and the locals still circulate in overcoats and woolen hats. Even as the days grow long, with the sun setting after ten o'clock, the trees in the parks and along the streets wear only a patina of lime-green fuzz. But things change fast. Barely a week later, Helsinki feels like summer. Along the city's posh Esplanadi, where twin boulevards flank a long park, sidewalk cafés are packed, and sun-starved Finns lie spread-eagled on the grass, soaking up the sunshine.

On a bright sunny Monday, Sampo Hietenan rides the elevator down from his midtown offices for lunch at an Asian fusion restaurant on the ground floor. He's wearing shorts and a striped short-sleeve shirt. Like just about everyone else, he spent most of the weekend outside. His face looks burned.

While Sonja Heikkilä dreamed up MaaS, Hietenan is among the first to build it into a business. His start-up, MaaS Global, sells monthly mobility subscriptions to a service called Whim. The promise, which is not yet fleshed out, is that subscribers will be able to name a destination and an arrival time. The app will promptly stitch together an itinerary, often with a choice of more than one conveyance. For some, it might involve a ride-share to a tram line and then a dockless bicycle to the destination. The fares for all of it will be included in the monthly subscription. Some users will pay more for richer blends of mobility—more taxi rides, less biking. Others will opt for economy packages, with larger doses of mass transit.

To date, the service is a work in progress. Most of the subscribers are on the "free" option. They use the app just to line up their transit but pay for each leg as they go. But the subscription side of the business is picking up steam.

Hietenan, in his early forties, still has a boyish look. While Sonja Heikkilä was putting together her thesis on subscription mobility, Hietenan was thinking along the same lines. He followed her work, like everyone else in the small mobility crowd in Helsinki, and then set about building it into a business. He brushes aside a shock of hair from his forehead, and leaving his salmon salad virtually untouched, he details in fluent American English the world-changing potential of mobility as a service.

"There hasn't been a productivity leap in transportation for almost a century," he says. Now, armed with smartphones and cloud computing, we're ready for the next one. He predicts that MaaS subscribers will save thousands of euros, while freeing up big chunks of transport time. "What would you do with another ninety minutes every day?"

The greatest challenge facing this new industry, Hietenan says, is that of dethroning the formidable champion, the automobile. "It's been around for a century," he says. "It might be the most successful business ever." And despite its significant downside—the rush hour gridlock, deadly crashes, smog, speed traps, and parking woes, among others—the car maintains its enduring appeal. People want a car, he says, because it represents the dream of freedom. You want to go someplace. The car is on call, that very minute, to carry you there. It's the closest humanity has come to weaving a magic carpet.

The car's domination isn't based entirely on its peerless range and availability. These machines are also tangled up with a host of our most primitive impulses and appetites. For a full century, some of the greatest minds in design, psychology, and marketing have worked to tighten the emotional bonds between ourselves and our machines. So naturally, in the middle of a chaotic city, many of us take refuge inside them. They provide a personal space, a cocoon. Further, in much of the world, the very buildings and cities we live in have been designed around the automobile. That is to say, most of us live in a car world.

How could a smartphone app ever compete? In these early days, it cannot. Simple as that. Even Hietenan and his wife still keep a car on hand, mostly to ferry their four children around. The first wave of his subscribers comes mostly from the carless minority. These customers range from cycling diehards, cash-poor students, and committed environmentalists to those who cannot afford cars (but *do* have a smartphone). Naturally, Sonja Heikkilä is a subscriber.

However, in a prosperous country like Finland, a mobility subscription service cannot thrive on these fringe groups alone. For MaaS to transform movement in Helsinki, and for the economics to work, it must reach a mass market. More passengers generate more supply, and only with more supply of everything, from trams to car-shares, can the city create a vibrant mobility ecosystem, one in which various mobility apps can face off, each of them offering enough options to compete with cars. To build that mass market, this nascent service industry requires economies of scale. This means appealing to a sizable slug of Helsinki's 300,000 motorists, coaxing them away from their hulking machines.

This poses a chicken-versus-egg problem. The service won't be great until car owners make the switch. Yet why would they switch before apps like Hietenan's can at least provide a service to rival their magic carpets? What could make them take the leap? One option, common in Internet businesses, is to offer the service at a bargain price, losing money to build up market share. That was Spotify's strategy in music and Amazon's in e-commerce—and, indeed, it has been at the core of Uber's strategy to carve out market dominance among ride-hailing apps.

But Hietenan believes that economics alone won't sway the motorists. Even if they commute on public transit to work, as many do in Helsinki, they keep their cars for the six or seven trips they take during the year to lake and pond cottages. It's a fundamental of Finnish life. This is the mind-set of car-clingers. These people—who represent about two billion of us, and most of the adults in the developed economies—often run scary simulations in the mind. They all have the same plot. You need a car for something, badly, and it's *not there.*

So people hang on to their cars, sometimes for the silliest

reasons. People who haven't played golf for two or three years still wonder how they'll carry their clubs to the course should the opportunity arise. In Japan, Hietenan says, nearly half the car-owning population drives less than once a month. They keep the car for that elusive weekend in the mountains, or perhaps an excursion to outlet malls. The car, he says, is an insurance policy. You pay more than it's worth—nearly 10,000 euros per year, on average, in high-tax Finland—for the assurance that it'll be there when and if you need it.

To prod people toward taking the leap from ownership to subscriptions, Hietenan plans to promise them magic. "What we have to do," he says, "is give them a dream." The MaaS subscription must tempt them with new experiences.

A hypothetical: Take a commuter named Pekka who lives in Espoo, the sprawling suburb across the bay from Helsinki (best known as the headquarters of the fallen cell phone powerhouse Nokia). A gleaming new metro line links Espoo to the capital, but Pekka lives a few miles from it, and he still prefers to drive his five-year-old Toyota Camry into town. (It helps that his company, like many in Helsinki, pays for parking.)

Let's consider Pekka's Camry. It certainly provides his personal space, or cocoon. Pekka can play loud music as he crawls with rush hour traffic and sing along as loudly as he wants. While few would call this a magic-carpet experience, the car at least leaves on his schedule. That said, the Camry, with its dirty cup holder and the car seat in the back, is short on the forces that might stir his animal blood. It's middling on power, and it offers little to nothing in the way of status. The carpet's caked with mud. As far as sex appeal goes, Pekka's commuting experience doesn't move the needle. His Camry, to return to RJ Scaringe's sartorial analogy, is the single outfit he's stuck wearing every day for five years.

"Now," says Hietenan, "how about if once a month we offer him a [rented] Ferrari, or a Porsche?" These types of deals will be possible as the legacy auto industry, from manufacturers to rental giants like Hertz and Budget, adapts to the service economy. They'll each devote a portion of their fleet to the MaaS ecosystem, he says, by the day or even the hour. Conceivably, a few hours in a luxury car would inject a dose of magic that, for the vast majority, car ownership never provides. Pekka might pick up his wife in the Ferrari for a Saturday night, giving them both a special experience. Or he could pull up to his Sunday-morning soccer game in a car that startles his teammates. Status!

Possibilities abound. It might be a couple of trips per month in vertical takeoff air taxis, which could be zipping over Helsinki within the next decade. Or perhaps a ninety-minute ocean cruise from the Helsinki harbor to the Estonian capital of Tallinn, including a flute or two of champagne on the return trip. Maybe only a tiny minority will make time for such a trip. So much the better, according to Hietenan. "If you find something they don't do, make it limitless," he says with a laugh.

Sampo Hietenan's dreams stretch far beyond Helsinki and Espoo. Once cities around the world implement their own mobility subscriptions, service providers can begin to offer roaming. A Finn with a roaming subscription could fly to Vienna or Los Angeles and then travel in those cities with the same multimodal magic, boarding buses and climbing into autonomous cabs, without worrying once about payment. In time, Hietenan says, a MaaS subscription could even cover air travel. A subscription would then be a ticket for a person's movement on the entire planet. That's why he named his company MaaS Global.

But before Hietenan can entice subscribers with Italian sports cars or seamless transit experiences in Shanghai, he has to put together a workable offering in Helsinki. The first step is to get all the transit data on the same standard, so that everything can work together on the same app.

Here he got a crucial boost from the Finnish government. In 2018, it passed legislation opening up mobility data to all comers. To do business in the country, car-shares like Uber and taxis alike must provide access to all their data—such as the location and availability of their cars, and all their pricing and timing information. This data allows vendors like Hietenan to offer it to their subscribers. Every mobility service, from the Helsinki subway to dockless scooters, must do the same. This legislation moves Finland decisively toward Sonja Heikkilä's original vision. All transportation options, by law, must adapt common open standards and mingle on the same apps.

Finland has some experience when it comes to unified tech standards. They've been a key to the country's national development strategy. In 1991, it was a Finnish company, Radiolinja, that launched the world's first digital cell phone network. The Finns later pushed for that technology, known as GSM, to become the European standard. The United States at the time had a grab bag of cellular technologies, both analog and digital, which often couldn't talk to one another. We take it for granted today, but in the 1990s, it seemed almost magical that Finns or Germans could take their phones to London or Lisbon, and they would *work*. This European standard gave small countries like Finland a true continental market and it helped launch a Finnish company, Nokia, into the world leadership of wireless communications. Finland, around the year 2000, was the unlikeliest of technology hot spots.

As long as cell phones were used mostly for talking, Nokia dominated the industry. But when it came to the next stage of mobile devices—smartphones—the Finnish company lost out. Apple's iPhone, released in 2007, established a new standard, and Nokia never caught up. In 2014, the company, which had once been worth a quarter of a trillion dollars, sold its cell phone business to Microsoft for a mere $7.2 billion. Within two years, Microsoft shuttered the division and wrote off the purchase price as a loss.

Not a happy ending for the Finns, but consider what their phone strategy accomplished. Their small country, with its population of 5.5 million (and a native non-Indo-European language inscrutable to 99.9 percent of humanity), became a leader, albeit briefly, in perhaps the most important technology revolution of the time, mobile communications. This brought billions in investments into the country and trained a world-class cadre of Finnish engineers and technicians. That brainpower still fuels the Finnish economy, and Finland's communications infrastructure places it among the global elite.

The question now is whether the Finns can pull off a similar trick in mobility. The strategy has some parallels. Starting with Helsinki, Finland can develop the model for the rest of the world and then export its mobility expertise and software. Companies like Hietenan's can expand beyond the country's borders.

But the far bigger benefit would be the transformation of Helsinki. If mobility as a service makes good on its promise, the Finnish capital could be the envy of the world: innovative, efficient, safer, and greener (though, admittedly, still a dark slog in the long winter).

Hietenan argues that the transition from the ownership economy could also free up tremendous wealth. He looks at the reigning car-centric status quo from the perspective of an investor. "What if somebody offered you a chance to invest in a factory that is idle ninety-five percent of the time?" he asks. "It would be absurd!" In his plan, the capital machinery of mobility would operate at multiples of this efficiency—with society reaping the economic rewards.

Once hundreds of thousands of Finnish car owners stop spending the 10,000 euros a year to buy, maintain, insure, and operate a machine that lies dormant for most of its existence, it will be as though each of these former car owners has gotten a big raise.

Part of that windfall, but only a fraction of it, will fund mobility subscriptions. But such services will be much cheaper by then, in large part because of superior resource management. Conveyances of every size and shape will be kept busy for many hours a day, generating a far greater return for society on the investment. Because of this, Hietenan reasons, a gold-standard mobility subscription—one with a car for cottage trips on demand, and maybe even that champagne-soaked cruise to Estonia—might cost only 500 euros a month, or 6,000 a year. Other subscriptions will cost half of that, or even less. That would leave all those former car owners with thousands of euros a year to spend on—whatever they want!

THE POSTCAR ECONOMY can be a lot of fun. Grab a dockless scooter in Austin, Texas, and ride it up to the university. Climb on a free metro in Tallinn, glide in an air taxi past the towering Burj Khalifa in Dubai. What's not to like?

However, in many cities, the push toward this transition involves not only dangling carrots, but also prodding with sharp sticks. In many cities, administering pain—making it harder and more expensive to drive—is a crucial element. London was a leader with its punishing anticongestion fees. It made it clear that driving in a city is not a right but a privilege, and an expensive one. This has been true since the first Model Ts, but gradually cities are catching on.

The even greater pain for drivers, the one that could shift the automobile age into reverse, is to cut back on parking. That hits motorists where it hurts, because the vast majority of personally owned cars need parking space just as badly as fuel. In Helsinki, parking lots under buildings represent an entire underworld dedicated to the automobile. This is mandated, in many cases, by the government.

Lots of people in the city government like to drive, and they worry greatly about reducing parking—and not only for themselves. It's a red-button issue for many of their constituents. They hear about it constantly. It's precisely the urgency of the parking issue that has led cities around the world to subsidize it, making it free on many streets, and devoting to it endless acreages of prime real estate.

In the Helsinki that Sonja Heikkilä imagined and Sampo Hietenan is working to create, this twentieth-century status quo cannot stand. In the Finnish capital, like every other city facing the mobility revolution, parking is where the policy battle begins.

Often it pits developers against city regulators. Finland's leading builder, YIT Corporation, recently applied to build high-rise housing near the port, but with reduced space for parking. The company, says Juha Kostiainen, the executive vice president for urban development, argued that parking spaces add needless expense. Already a lot of young people are moving around without cars. More of them would be able to afford downtown units if they didn't have to pay for the parking. Each underground space costs about 55,000 euros—well over the 40,000 euro average annual income in Finland. "We said, 'We'll take the risk,'" Kostiainen says. "'Let us build it with fewer spaces. If it's not popular, the market will drive down the price.'"

The idea didn't sell at city hall. The fear, of course, is that if city officials cut back on parking, legions of angry motorists will endlessly circulate in Helsinki, fouling the air and congesting the streets in the hunt for parking spaces that no longer exist. Politicians aren't blind to the future. But if they impose it on motorists today, they fear, frustrated drivers will vote them out of office.

It's a crucial policy question in cities around the world. In Helsinki, it pits the environmentalists, or Greens, against conservatives, city dwellers against suburbanites. The central question is whether cutting back on parking will spark a political backlash. Or could it change people's behavior? If parking becomes too much of a headache, will more motorists stop driving to town, with some of them even dumping their cars and subscribing to a mobility service?

There's no doubt where Otso Kivekäs stands on the issue. He's an activist in the local Green Party and the president of the Finnish Bicycle Association. He works in a ramshackle office in an old Helsinki building with high ceilings and peeling paint. Kivekäs has the sturdy build of a cyclist who pedals through the long Helsinki winter. He has a black beard and a long ponytail. A tattoo of a blue snake winds down his left arm.

Otso Kivekäs and his wife have no car, and he can drag their two young daughters on a wagon attached to his Dutch-built electric bike. He and his fellow Greens, backed up by seventeen thousand activists on their Facebook page, successfully pushed the city to clear central bike routes through the winter. This isn't a matter of just plowing the routes. That leaves a dangerous layer of ice. And forget about spreading salt on the trails. That busts tires. Instead, Helsinki's transit workers sweep away the snow with long brooms. It's painstaking work.

In the battle for the future of Helsinki, Kivekäs is a hardliner. As he sees it, a war is raging for space in the city, with cars on one side, and all the other transit on the other. The space itself is immutable. It can present either a welcoming environment for automobiles, or—Kivekäs's strong preference—a hostile one.

Outside Helsinki's small and walkable downtown, Kivekäs

says, the city was built for cars. Its sprawl makes it a bit more like an American city than neighboring Stockholm or Oslo (though no one will ever confuse Helsinki's sprawl with Houston's or LA's). He says that planners through the decades imposed single-use zoning regulations, so that businesses and schools and hospitals and houses were all separated from each other, and then stitched together with highways. "Everything the planners have done, for decades, has been wrong," he says. The goal of the Greens is to add density to the metropolitan area, reducing roadways and packing in more housing and commercial space, so that more people live and work within biking distance or a short walk from a subway or tram.

Naturally, this is a hard sell to the car-centric members of the Conservative Party. But the Greens, Kivekäs says, have found a way to split their opponents. It has to do with money—specifically, the money to fund the kinds of multiuse construction projects the Greens want. "The Conservatives have two wings," he says with a broad smile, "the car wing and the construction wing." Cars bring votes, but construction brings money. Backfilling Helsinki's sprawl and preparing the area for the next stage of mobility involves lots of construction contracts—which can gain bipartisan support.

The result is a city plan heavily influenced by the Greens. Seven highways connect Helsinki to the suburbs. The new plan would bulldoze these broad highways and convert them into narrower and slower surface roads. It would then reclaim much of their multilane footprint for sidewalks, bike lanes, green medians, and tramways.

That may be a bit too much for the federal government, which has to consider not only a motorists' revolt but also truck traffic to the port—Finland's crucial shipping link to the rest of Europe. The result may be that the Green plan gets dialed back, with perhaps only two or three highways turned into greenways. Still, while too slow for anticar radicals like Otso Kivekäs, the conversion is moving ahead. Helsinki is molding itself to the next generation of mobility. That brings Sonja Heikkilä's dream of mobility as a service closer to reality.

THE RECEPTIONIST PLACES a call, and a few minutes later a tall young woman with long blond hair appears in the lobby of OP Financial, Finland's biggest bank. She wears a lustrous cobalt-blue dress.

Once we're seated, Sonja Heikkilä explains, in her meticulously correct English, how she came to be a banker. After finishing her degree, she worked at Tekes, the government's funding arm for innovation and technology. She focused, naturally, on putting together the pieces for the mobility revolution.

Officials at OP Financial, meanwhile, saw that the revolution Heikkilä was helping to orchestrate would likely disrupt all kinds of businesses, including their own. The bank historically had 10 percent of its business in autos, both financing and insuring them. What was going to happen to those revenue streams, it wondered, if hundreds of thousands of Finns stopped owning cars and began subscribing to mobility services?

They brought Heikkilä on board in 2016 and gave her a simple mandate: to come up with new banking services for the next stage of mobility. As far as Heikkilä is concerned, she hasn't strayed from her goal to revamp mobility in Finland. "But now," she says, "I have the resources of a big bank."

The paradox is that the person who envisioned the next stage of mobility is now devising new ways to provide cars to people. She maintains that in those heady student days, when she seemed to be the Finnish oracle of mobility, many misunderstood her goals. "They thought I wanted a world without cars," she says.

"But what I wanted was a world where people don't *need* to own cars."

Sonja Heikkilä still doesn't own a car, and she subscribes to Sampo Hietenan's mobility app, Whim. But she knows from her own experience that the app still has big holes in it—ones that, for the time being, only cars can fill. She needs a car, for example, when vacationing with her family on the coast, or to lug her athletic gear across town. Cars, despite their shortcomings, are still unrivaled for getting to lots of places and moving stuff.

The car isn't the enemy, in Heikkilä's view. The enemy is the inefficiency of car ownership, where the resource sits idle for hours on end, hogs precious space and, more often than not, moves only one person at a time. The key for the bank, she says, is to come up with a host of flexible services—she has five, and counting—that provide cleaner electric cars to individuals and companies for the hours and minutes they need them. In this scheme, fewer Finns buy cars, and the ones that circulate are used closer to their capacity. It's less wasteful.

It's just a matter of time before car services like Heikkilä's get packaged into the mobility subscriptions that Sampo Hietenan is selling. It's not the future without cars. They'll still be around, but there will be fewer of them. Instead of the titan of the mobility economy, cars will be just one more rolling piece of it.

6

In the Company of Hawks and Nightcrawlers

If you drive up Mexico City's Paseo de la Reforma, past the golden statue of the Angel of Independence, through Chapultepec Park, and continue uphill for a couple of miles, you'll find yourself in an exclusive neighborhood of stately mansions, some of them with armed guards at the gate. This is Lomas de Chapultepec.

A canopy of oak and jacaranda trees blocks most of the view. But from certain spots, if you look back, you can see the immense city below. It stretches to the horizon, the pastel pinks, blues, and greens of the houses muted on most days by a blanket of smog. The roads of the capital, the jammed *avenidas* and *calles*, look like the gray lines of an Etch A Sketch.

A good distance back, on the far side of the central Zócalo and the sixteenth-century cathedral, is Mexico City's Benito Juárez International Airport—a key destination for many in Lomas de Chapultepec and the other wealthy enclaves on the neighboring hills. It's not a pleasant drive to the airport, to put it mildly, or a fast one. At rush hour, the twelve or thirteen miles can take an hour and a half. If there's a crash on the Viaducto Miguel Alemán or the Circuito Interior, it might drag on twice as long. When traffic is paralyzed, the threat of street crime looms, especially for those in luxury cars. Some of Mexico's rich travel through the city with bodyguards.

About five minutes downhill from Lomas de Chapultepec, a company called Voom, a subsidiary of Airbus, runs a heliport. Voom offers a ten-minute flight to the Mexico City airport, or a longer one to the airport in neighboring Toluca, for about $140. The company also sells helicopter service in São Paulo, a mega-city where the dynamics of wealth, traffic, and crime are sadly similar.

At first glance, there's little new here. Rich people spend more than a laborer makes in a week to save an hour or two. They fly over the hazards and headaches that everyone else is stuck with. Such are the perks of wealth.

But what if smaller heliports popped up all over the metropolitan area and offered flights on petite and quiet electric machines at a fraction of that price? What if everyday people could call up a flight on a ride-sharing app? That's the vision at Voom, and at a host of other flight companies around the world. The future, as they see it, features much smaller electric vertical take-off and landing airships, eVTOLs. Uber has announced plans to launch networks of these flying machines in Los Angeles and Dallas by 2023. A host of companies, including subsidiaries of Airbus and Boeing, and Germany's Volocopter, are angling for the first contract in Dubai, which might come even sooner.

In a promotional video for Uber, a passenger strolls to a rooftop launch area, where she is escorted with other lucky folks onto the Airbus-manufactured vessel. Later she looks out her window at the usual snarls of traffic below. She shakes her head, clearly relieved to have advanced to a higher level of mobility.

Yet on the sky side of the mobility revolution, crucial questions remain. It's not just whether the flying technologies will work—we can see from the plethora of drones that most of them already do, at least in controlled settings. The bigger issues are whether they can spawn profitable and sustainable businesses, and whether society will manage and regulate them intelligently, for safety, the environment, and social equity.

The normal expectation, given the history of the helicopter and private jets, is that the skies will remain the preserve of the rich. However, a business for the top 1 percent would mark failure, because the case for eVTOLs is based on mass markets—busy networks ferrying around large numbers of people.

This is impossible to achieve without advancing beyond the helicopter. If you're riding across Mexico City or Midtown Manhattan in one of them, it's a dream, albeit a loud one. But much about the technology is pretty awful. They guzzle fuel and require skilled pilots. When landing and taking off, the propellers stir up thick clouds of dirt and turn pebbles into stinging BBs. The experience is deafening. Fabien Nestmann, head of public affairs at Volocopter, calls noise the biggest impediment. "No one wants to have a helicopter landing next to them."

IT WAS NAMED the Puffin, after the burly northern seabird with the bright-colored beak. In 2010, NASA rolled out a prototype of this one-person electric flying machine at its Langley Research Center in Virginia. Like its namesake, the craft was small—barely big enough for one horizontal passenger—and it had short wings. It was also fast, with projected speed reaching 150 miles per hour. When NASA posted an animation on YouTube of a Puffin prototype, the video went viral.

The Puffin's creator was a thirty-year NASA researcher named Mark Moore. The affable Moore, who fittingly wears aviator glasses, had been developing designs for advanced aviation at NASA. This was a bit of a backwater at the Langley lab. From NASA's perspective, oddly enough, while unmanned missions to

Mars were real and fully funded, the vision for small and autonomous aircraft flying over major cities existed largely in the realm of concept.

As electric technology and computers advanced late in the first decade of the twentieth century, and the automaker Tesla took off as a commercial success, the idea of electric urban flying machines stopped sounding quite so outlandish. In 2010, the same year that the Puffin prototype was unveiled, Moore published a white paper mapping out the technical path for small eVTOLs. Clearly they were within reach, perhaps within a decade. A nascent industry took notice. After reading Moore's paper, according to *Bloomberg Businessweek*, Google's CEO, Larry Page, launched two aviation start-ups. Boeing and Airbus funded their own efforts. A global race was under way.

Today, some one hundred different vehicles are in development around the world. This stampede came about virtually overnight. François Chopard, whose Franco-American tech incubator, Starburst Accelerator, is invested in the sector, describes it with a tone of wonder. In the summer of 2017, he says, there were just a few corporate players, like Airbus, and a handful of start-ups. "A year later, we were able to count eighty start-ups." He compares it to the first age of aviation. By 1909, just a year after Wilbur Wright's first public demonstration of a manned aircraft in Le Mans, France, fifteen aviation start-ups had popped up in the Paris region alone.

Hype is thick and pervasive, but the explosion of activity and investment is real. And if you look at the evolution of electric aviation, hoisting people around is the logical next step. In its early days, starting in the 1960s, most of the electric flying machines were toys. Kids got the remote-control gadgets for the holidays (and usually crashed or lost them by New Year's). In recent years, infinitely more sophisticated versions of that technology, packed with digital smarts, have evolved as drones. Now it's a matter of preparing those vessels for passengers, adding lots of redundant propellers and power—just in case—and operating fleets of them in transit networks.

They come in every shape and configuration, some solely with propellers, others combining them with fixed wings. The Volocopter, which has already flown autonomously and with great fanfare across Dubai, lifts off like a horsefly, powered by a circular crown fitted with eighteen whirring propellers. In New Zealand, one of the companies Google's Page funded, Kitty Hawk, has been testing a craft called Cora. It looks like a World War II–vintage British Spitfire, but with three electric propellers on each wing.

In these early days, virtually everything gets funded, and anything goes—at least until the winners emerge. "What you see now is a genetic algorithm playing out in real time," says Brian Yutko, vice president for research and development at Aurora Flight Sciences, a Boeing company. From wing and battery design to propulsion, Yutko says, "there are dominant features that are starting to form. They have arms and legs and are crawling out of the swamp."

Yutko describes this Darwinian march at a 2018 roundtable sponsored by Uber Elevate. The host of the panel, the director of aviation at Uber Air, is none other than Mark Moore. In 2017, as he closed in on full retirement benefits at NASA, Moore quit to join Uber. This would be his chance to turn his research from concept to reality—to help develop an entirely new industry of mass mobility in the air.

Moore's vision is anything but modest. He foresees what he calls "dynamic skyline networks" flying hundreds of thousands of people a day in major cities within a decade, and eventually "millions of trips a day for a city like LA." He predicts that the cost of such air transport will plunge to the marginal cost of a simple car trip.

Given the technical challenges still ahead, in battery technology and AI, the timing on Moore's forecast may well be a bit rosy. And the scheduled launch of Uber Elevate services in 2023 is no doubt a push goal. Moore admits as much. But he argues that a tight deadline from an ambitious and deep-pocketed player like Uber has goosed the entire industry, unleashing billions for research and development of the technology. A bit of hype, used judiciously, isn't always such a bad thing.

WHEN ENVISIONING A new technology, the challenge is to free our imagination from the reigning status quo. It's easy to think about an eVTOL as simply a smaller and cheaper helicopter, one that happens to be electric and, eventually, self-flying. In this thinking, the flights from the heliport near Lomas de Chapultepec would continue service to the airports, but maybe they'd charge lower fares. Since the eVTOLs would be so much quieter, and less obtrusive, maybe they could fly other routes, too, from the wealthy Polanco neighborhood, or perhaps from Cuernavaca in the next valley to the south. But that's helicopter thinking.

While most passenger helicopters fly from point A to point B, eVTOLs would function as urban networks. As Moore describes it, a city would sprout dozens of small air stations, similar in scale to elevated subway stops. People would fly from one station to another, perhaps making a stop en route to pick up another passenger or drop one off. Different eVTOL operators could compete on the same network, just the way buses and cabs compete on the roadways. But here are the fundamental differences: If a subway with fifty stations adds a fifty-first, one of the lines grows a bit. But when such an air network adds a station, fifty new routes materialize. It's a vision of mass transit above our heads.

Not everyone is crazy about the idea. Mobility technologies, after all, have a way of taking over. Early in the twentieth century, when the first automobiles were circulating on city streets, few could have imagined that within decades they would colonize urban geography, paving and reshaping it to their special needs. If a big chunk of urban transit moves into the air, won't it blacken our skies?

Moore insists that it won't. "Don't imagine that Star Wars sky," he says, "with thousands of starships flying over your head." Most people, he says, will barely see or hear the airships. They'll be small, closer in scale to eagles than jets, and their whirring noise will blend into the dull roar of a city's soundscape. Such assurances, especially coming from a company building a business in the sky, won't satisfy everyone.

To address safety concerns, especially regarding autonomous flight, most services will launch their fleets with human pilots. But the plan is for the software to take over toward the end of the decade. Autonomy, in fact, is central to the business model, a key to building air taxis into a mass market. "To get the economics to where you need it to be, you have to get the pilot out," Ryan Doss, the head of product at Karem Aircraft, a Southern California start-up, tells the crowd at an Uber Elevate conference. What's more, there simply aren't enough pilots to run large networks of eVTOLs in leading cities around the world. "It would be an incredible challenge to train up a whole bunch of pilots," Doss says.

While automation is new in cars, it's old hat in aviation. Even before World War II, airplane manufacturers were introducing cruise control functions. This would have been unthinkable on the confusing thicket of roads on the earth's surface. But it made sense in the open skies. Since then, the industry has added layer upon layer of autonomy. It enhances safety and relieves the pilots of their most tedious work, such as maintaining a constant speed, altitude, and direction. Glitches do occur. Two crashes of Boeing 737 Max airliners in late 2018 and early 2019, in Indonesia and Ethiopia, appeared to involve software problems. Nonetheless, flying in commercial liners remains by far the safest way to travel, and automation plays an important role in that.

Modern passenger planes operate at what for the auto industry would be level 3 autonomy, similar to Tesla's electric cars on autopilot. A human being must be at the controls, but he or she doesn't have to do much. Like autonomous cars, planes are fitted with dozens of sensors, to measure the wind, air pressure, temperature, humidity, turbulence, and every other quantifiable variable. A computer on board crunches all that information—more than a petabyte of data for most flights—and handles the lion's share of the flying by itself. On most trips, the human pilot takes off, and he or she then turns the flight over to the computer, which can maintain a steady course and altitude, and respond to changing conditions. In many cases, the computer even handles the landing, before the pilot takes back control to taxi to the terminal.

Urban flight is much simpler. While jetliners, flying at 35,000 feet, grapple with jet streams and wind shears, eVTOLs take tiny hops within a couple thousand feet of the earth's surface, just clearing the tallest skyscrapers. In a nautical parallel, the big planes are like ocean liners braving the open seas, while eVTOLs are more like speedboats zipping around the marina.

Perhaps the biggest challenge for eVTOLs is to stay clear of each other. For this, they'll run constant network connections with one another, sharing their speed, position, and planned route. The goal is to have them as coordinated and aware of each other as a murmuration of starlings rising and dipping in time.

The software challenges, for both the network control and

the automation in the plane, are significant. Engineers have to ensure safety even when things go wrong—when engines fail, propellers break, storms erupt, or the vessel is convulsed by an electromagnetic charge from a cell tower below. A success rate of 99.9 percent isn't even close to good enough.

Still, autonomy will be essential for safety. To understand why, put yourself in the swivel chair of an air network administrator in near-future Los Angeles or Dubai. It's like air traffic control, but on steroids. On a screen, you might see hundreds of flights run by competing services zipping among dozens of air stations. These are the Ubers and Lyfts of air travel. Now imagine that these flights are being piloted by human beings, all of them figuring out for themselves their ideal speed and direction. How could you possibly manage all those pilots and keep them out of each other's way?

Even if each pilot pays zealous attention to the stream of commands from the control center, precious seconds are lost as the order gets communicated to the human brain, processed, and transmitted into physical movements. While it's true that traditional air control routinely manages steady streams of takeoffs and landings at major airports, scaling such systems to manage hundreds of flying taxis would be impractical and dangerous.

What this network administrator needs is a computer that micromanages the flight routes, second by second, and an entire fleet that follows commands with the precision of . . . machines. It's an industry that cannot thrive, or even sustain itself, without automation.

This doesn't mean it will be an easy sell. The networks must be fast and secure. The eVTOLs must communicate not only with central command but also with each other. And they must calculate the most efficient routing. If most of the rush hour flights go downtown, how do the eVTOLs redeploy, but with a minimum of empty flights—known as "deadheads"? Will they offer sharp discounts, or perhaps air-deliver packages or emergency plasma? Success hinges on inventory management and dreaming up new possibilities for urban air markets.

Powerful batteries are also critical. For air taxi services to work at scale, the machines will need to be up in the air and making money for many hours a day. Every hour they spend recharging their batteries will lower their yield. "I believe we'll get batteries that are powerful enough," says Ahmed Bahrozyan, the chief executive of public transport at Dubai's Roads and Transport Authority (RTA). "But I don't know when."

Meantime, municipal governments will come with a host of legitimate concerns. First and foremost: What happens if a flying machine falls from the sky? It's true that cars crash, and that's bad. But they don't land like bombs on sidewalks, apartment buildings, or, heaven forbid, playgrounds.

The industry line—that the machines will be quiet and virtually invisible—will find few believers among a skeptical public. Opponents will no doubt warn of whirring airships dominating the skyscape, an infestation. Many city dwellers already gripe about the proliferation of electric scooters on the streets and sidewalks. They're not likely to feel generous when it comes to the air above.

Of course they'll pose legitimate questions about safety, and they'll likely blend them with a strong dose of class warfare. After all, the very first patrons of eVTOL services are likely to come from the ranks of helicopter riders, the Lomas de Chapultepec crowd. Air service from fancy neighborhoods to airports and business hubs is the low-hanging fruit. Not everyone will be thrilled.

All the mobility technologies we've been exploring in this book, from scooters to autonomous cars, raise thorny issues for municipal governments and regulators. But none of them comes close to the challenge of flying taxis. In the United States, for example, the Federal Aviation Administration supervises the pathways that carry air traffic to and from airports. But whether an eVTOL is landing at Dodger Stadium or crashing into a backyard in Queens, it's ruled and regulated by a city government. "The FAA makes them fly," says Los Angeles mayor Eric Garcetti. "We make them land."

Establishing the rules to govern this new branch of air traffic will be a grueling job for city governments. It will require safety regulations, technology certification and licensing, fee structures, and cybersafety guarantees. (Hackers could inflict massive harm, conceivably even turning a fleet of eVTOLs into weapons, much the way al-Qaeda commandeered jetliners for the September 2001 attacks.) City governments might also push carriers to ensure equity, providing affordable access across an entire city. The whole process is bound to spawn heated debates, and in some places, angry protests.

Yet governments dealt with similar issues a century ago, at the dawn of the aviation age. As authorities haggled over rules, city residents, especially those near the new airports, worried about their safety and the noise. Newspaper editors wondered why people had to be in such a hurry. Was it really necessary?

The industry's first job was carrying the mail. In that sense, it was nurtured by the government, and it was able to establish safety standards before private airlines launched their passenger business. Once the public got a taste for air traffic, it quickly became an addiction. In 1926, there were fewer than 6,000 airline passengers in the United States, but within three years, traffic had ballooned to more than 170,000. By the end of the 1930s, a million passengers were flying in the United States.

Over the following decades, the geography of our economies

was marked, in great part, by airports. Cities like Atlanta, Dallas, and Dubai would be small dots on the map without their giant air hubs. At the same time, we retuned our sense of time and space to the jumbo jet. Paris was six hours from New York. London, five. We flew, because it was faster and worked better.

"Faster and better." It has a strong appeal. That's why cities like Los Angeles and Dallas, and a host of others, are so eager for these tiny sky ships that they're ready to take on the thorny issues. They'll meet with technologists and insurance companies, and they'll discuss the risk of falling machines. They'll hammer out regulations with lawyers and state and federal officials. They do all this, in great part, because they're desperate for relief from knotted traffic. A 2017 survey from transportation analysis firm INRIX listed Los Angeles as the most congested city in America. Dallas rated twelfth. Like Mexico and São Paulo, these cities will no doubt look for eVTOL relief first in rich enclaves, places like Bel-Air and Dallas's Turtle Creek, before dotting their geographies with air stations.

IT SOUNDS UNLIKELY, in the near term, that an air taxi service would reach a scale sufficient to improve the flow of ground traffic in a major city. Commuters in the world's big cities number in the millions, and most of them drive cars. Could an air service make a dent in that?

For decades, mathematicians and engineers have been meeting at conferences and colloquiums to discuss the dynamics of traffic. They never seem to reach a consensus. What causes so-called ghost jams, for example? That's when you're driving down a highway at normal speed, and next thing you know, traffic is paralyzed. You creep forward for a few frustrating minutes and then, in an instant, find yourself traveling at full speed again.

You look for an accident or a problematic on-ramp. But the jam just seemed to happen, and then, just as quickly, vanish.

Some researchers theorize that the phenomenon follows the mathematics of shock wave propulsion. Drivers hit the brakes, maybe to rubberneck, and that deceleration moves backward in cascading waves, until all forward momentum is extinguished. Others compare the dynamics of car travel to that of viscous liquids such as honey. Theories abound. But one point of agreement is that traffic does not follow a linear model: If you take 10 percent of the cars off the highway, traffic does not speed up by the same amount. The impact might be much greater.

This is the point Mark Moore is eager to make. He argues that if air taxis bring about even a 2 or 3 percent drop in rush hour traffic, it could improve the flow by 10 or 20 percent.

However, if traffic theorists, using models of viscous liquids or shock waves, can make a convincing case for flying taxis, game theory might offer a counterargument. Planners, after all, must calculate not only the first-day impact of a new element, whether a new bridge, bike lane, or air service. They also have to anticipate how people will respond to it. If new air taxis reduce traffic jams by 10 to 20 percent, could it be that more drivers will be lured back from public transport to their cars? It's anybody's guess.

But even if the air taxis don't revolutionize traffic, they play to a key constituency: the powerful. Put yourself in the position of New York's mayor, Bill de Blasio. He meets with banking titans, film producers, and tech executives. These big shots could bring in millions of dollars in investment and create jobs. He wants them to pick New York and not London or LA. Yet if they fly into Newark or LaGuardia or, heaven forbid, JFK, many of them will have endured not only cramped and backward airports, but a miserable trek, in a cab or limo, all the way to Manhattan. It may be their freshest memory when they appear at his door—the worst welcome mat imaginable.

If these same people visit LA and there's a functioning air network, even in its infancy, it could be an entirely different experience. They zip from the airport to city hall in a quiet and affordable air vessel. Along the way, they hear that the service will expand to Santa Monica, Pasadena, Hollywood. This delivers a message that the city works, that if mobility is not yet a breeze, it soon will be. Even the promise of mobility can fuel growth.

The solution, it seems, is to free ourselves from the crowded surface of the earth. But not all the action is in the air.

ELON MUSK CERTAINLY has nothing against aviation. He's the founder of SpaceX, and he hopes one day to send spaceships all the way to Mars. Still, he ridicules the vision of the LA skies abuzz with swarms of flying cars. He imagines one of them in need of servicing. "It will drop a hubcap," he says, "and guillotine someone."

Musk has come to a traffic-jammed section of wealthy Bel-Air, where he's speaking to a synagogue congregation. His solution, he tells them, is not to look to the skies for traffic relief, but to the earth below.

This vision leads Musk to yet another mobility frontier, one beyond rockets and electric cars: earth-burrowing technology. His tunneling start-up, the Boring Company, is digging for riches in a realm long dominated by worms, moles, and sewage lines. The goal is to be zipping passengers through an underground labyrinth at 150 miles per hour. If this vision pans out, a twenty-mile ride from the airport to Dodger Stadium, which cars usually cover in about an hour, might take ten minutes.

When Musk launched the company in 2016, to bore a tunnel was an excruciating exercise in patience. Steve Davis, his chief operating officer, puts the inch-by-inch struggle into context. The strolling pace for a human being is about three miles per hour. Snails slide along on their carpet of goo at about 1/100 that speed. A state-of-the-art tunnel-boring machine moves ten times slower than a snail. Musk's initial order to his team was to speed up the hulking machine by a factor of ten—to match a snail's pace.

That goal has been achieved. The next one is to accelerate it by a factor of fifty, to the walking pace of a toddler. At a toddler's pace, it would take the company a mere two weeks to bore a 415-mile underground link between LA and San Francisco. That would enable a related dream of Musk's, Hyperloops between cities, which feature trains whistling through vacuum tubes, conceivably at supersonic speeds. (Airless tunnels, according to Davis, are impervious to the sonic booms created by sound waves.)

For the current project, the underground trains of LA (an electric but unvacuumed network known as the "Loop"), Musk envisions a vast underground warren of hardened pathways with electric-powered platforms, or "skates," moving passengers at up to 150 miles per hour. He hopes to build a first line from the airport to the Coliseum, downtown, and to Dodger Stadium, much of it along Interstate 110. A second route was envisioned to run forty miles under the hellish 405 freeway, from Sherman Oaks, north of downtown, all the way to the Long Beach airport. But suits filed by community groups led the Boring Company to shelve that plan.

If all goes well on the first routes—if the Loop works as designed—Musk won't be satisfied with just one level of tunnels under Angelenos' homes. No, that would limit growth to two dimensions, just like the jammed highways above. His goal is to create multiple dimensions of tunnels, perhaps as many as one hundred.

He also plans to design a transit experience that in many ways is closer to that of the car. The Loop, Musk hopes, will tend to the individual, and not the herds. Customers will not congregate at stations, because that creates congestion. In Musk's view, when people are at stations, they're not where they want to be. They still have to get somewhere. That's an immense inefficiency.

So unlike transit networks, with their stations and hubs, Musk is mapping out a scheme much like Mark Moore's aviation vision. It's a distributed network, in which software manages the traffic, dispatching perhaps tens of thousands of these electric pods like packets of data zipping through information networks. Eventually, Davis says, people could access these tunnels—pod ports—from their own homes, maybe climbing down from their carless garages.

The technical challenges ahead might not be as daunting as developing a shuttle to Mars. But for building a business on earth, it's a bear. Speeding up tunneling technology by a factor of fifty is alone a stretch goal. So is building and deploying an entirely new mode of transportation. But perhaps the biggest challenge involves public relations. To get the go-ahead to create his underground dream in Los Angeles, followed by other cities, Musk has to build trust and goodwill—which is precisely why he and his operations chief have come to the synagogue in Bel-Air.

The most frequent questions concern earthquakes. In fact, Musk insists, tunnels are the safest place to be in earthquakes, which jolt the earth much more on its surface than underneath it. Davis compares an earthquake to a tsunami, which can sink a heavy battleship while leaving unscathed a submarine operating near the ocean floor.

The other common concern is noise, but Musk says not to worry about it. The electric tunneling engines, powered by Tesla batteries, he says, make far less noise than the standard diesel version. So for humans at ground level, the work going on thirty feet below is inaudible. What's more, the tunneling takes place far below the electrical wires and water and sewage pipes that undergird the big city. At that depth, the big dirt-chugging machines don't even encounter wildlife. Even the hardiest worms, Musk says, stick closer to the surface.

The tunneling machines, each one named after a poem, are still digging at much closer to a snail's pace than the steps of a toddler. The biggest problem is disposing of the dirt. When you're thirty feet underground, where can you put it? Fifteen percent of a tunnel's cost, Musk says, is "getting rid of the muck." One of his company's answers is to compress the dirt at thousands of pounds, creating hardened bricks from it, and then using those bricks to reinforce his tunnels. "They're very beautiful bricks," he tells the audience in a quiet, reverential tone. They think he's joking and laugh. But he's perfectly serious. The bricks represent an early breakthrough in an ambitious process that's going to require many more of them.

7

Dubai: Grasping for the Cutting Edge

It was a sunny November afternoon in downtown Los Angeles, in the rapidly gentrifying Arts District. This was the first gathering of CoMotion LA, our annual conference of movers and shakers of the global mobility revolution. Along a closed-off alley lined by new mobility start-ups, attendees were zipping back and forth on the latest electric bikes and scooters and a host of other whimsical machines. A rotund robot on hidden wheels was following people around, as if drawn to them. It was a relaxed scene.

But toward the end of the day's session, a formal presence entered the picture. A middle-aged man, bald and wrapped in a dark business suit, made his way into the auditorium and walked deliberately to the lectern. He began delivering a speech in slow and laborious English. None of the usual anecdotes for this speaker, much less jokes. He clicked through PowerPoint slides. Some in the crowd no doubt suspected that one of the sponsor companies had wrangled a prime afternoon speaking slot. It happens.

The speaker was HE Mattar Al Tayer. The "HE" stood for "His Excellency." He represented the emirate of Dubai, and he headed up the powerful Roads and Transport Authority—the ministate's transportation ministry. While Al Tayer's delivery was dry, his words were anything but. Unlike most of the speakers at CoMotion LA, who described the coming mobility revolution or detailed the components they were building, Al Tayer and his team in Dubai were putting the revolution together—creating the future of mobility—and fast. He was talking specific products and services, and even *dates*.

Al Tayer didn't merely envision a future in which robotic air taxis lifted people above Dubai's growing expanse of skyscrapers and knotted highways. They would be buzzing the skies by 2020, he said. His government had already overseen a test of flying pods, made by Germany's Volocopter. A Hyperloop would be connecting Dubai with its neighboring emirate Abu Dhabi. The two-hour drive would shrink to fourteen minutes. Al Tayer also promised that within a decade, one-quarter of the mobility in Dubai, a fast-growing city of three million, would be autonomous.

On its patch of desert on the Persian Gulf, it appeared, Dubai was busy building the Jetsons' future.

As we put together ideas for this book, it was clear from day one that Dubai would anchor a chapter. Other cities dropped on and off our list, but Dubai was a constant. The fast-growing emirate represents a starkly different model than that of Los Angeles or Helsinki. Both of those cities are encouraging mobility revolutions by opening doors and tweaking regulations. They're inviting change, but they don't have the resources, much less the political authority, to build or mandate much of it.

Dubai's in a different boat. It has more in common, as we'll see, with China. Neither faces the messiness or tiresome delays of a democracy. New York, for example, dithered a half century on extending Manhattan's East Side subway by a measly three stops.

Consider what Dubai has pulled off in just twenty years. It

has conceived and built its signature skyscraper, the Burj Khalifa, as well as its metro. It's laid down countless roads and erected entire neighborhoods of high-rises. In terms of building, only the Chinese megacities operate on a similar massive scale. Projects in Dubai are not slowed down by petition drives or bruising battles in city council. The sheikh calls the shots. Autocracies, as their defenders have long maintained, are far more efficient, at least when they work. Dubai's, indubitably, works.

And Dubai has a government eager to spend billions to conjure its sparking mobility visions to life—as many of them as possible before Dubai hosts its 2020 expo, a grandiose exhibit of the future of human technologies and possibilities. However, Dubai is also very different from China, the other authoritarian state we'll be looking at. While China is pushing to dominate the coming technology revolution and overtake the United States as a tech powerhouse, Dubai, for the most part, is a mere practitioner of the cutting edge, a highly enthusiastic consumer. It looks to technology to help catapult its economy toward the future, and to bolster the shiny brand of Dubai Inc.

You might think that Dubai, part of the United Arab Emirates (UAE), would draw on rich oil reserves to underwrite new mobility. However, this is not the case. Unlike Abu Dhabi, its wealthy neighbor eighty miles to the south, Dubai is a relative pauper when it comes to energy. Its modest oil fields are largely spent. Instead, it makes its money as a lively hub for shipping, commerce, banking, and tourism. Perhaps the closest model to Dubai is Singapore, though as we'll see, it also has parallels with Las Vegas.

Dubai's economy is based on attracting people and money to its city. Its growth hinges on a steady stream of investors plowing their millions into the emirate. Its job is to draw them in, to entice them to keep building skyscrapers, hotels, even artificial islands shaped like trees. The foundation of Dubai's economy is real estate, and this requires a never-ending pitch.

That's where all the new mobility comes in. These projects, from the ever-expanding subway to the fleets of autonomous taxis, will no doubt enhance movement in the city. That's important. But they're also central to Dubai's glittering brand. To keep attracting tourists and investors, and to keep building, Dubai doesn't just want the hottest technologies. It wants them first. Dubai's ruler, Sheikh Mohammed bin Rashid Al Maktoum, declares as much in his 2012 book, *My Vision*: "We must take the lead in forging our own destiny," he writes. He argues that technology has unleashed "the most important economic race that the world has ever witnessed."

To someone who visited Dubai even a generation ago, the idea that this desert city would aspire to such greatness would seem almost laughable. But Dubai, like Shenzhen and Orlando, has utterly transformed itself within the lifetime of many of its citizens. In the 1930s and 1940s, it was a small trading outpost with a contingent of deep-lunged pearl divers. The city's sole asset was a deep inlet harbor. The tiny sheikhdom by the sea offered free commerce and low taxes, and a number of Iranian merchants crossed the gulf and set up shop there. In the early decades of the twentieth century, the Dubaians thrived, briefly, as the world market for pearls boomed. But luxury items plummet in global depressions, and the local pearl business withered in the 1930s. The surviving remnants were snuffed out by Japan's rising industry of cultured pearls. It was as a small port and trading post that Dubai eked out its existence. Dubaians wouldn't have electricity, much less air-conditioning, until the 1960s—the same decade in which they finally abolished slavery.

The emirates existed in a largely ignored corner of Arabia. The British fleet provided them with a measure of protection.

But as the fading empire retreated after World War II, the British loosened their commitments, and in 1971 they left the small Gulf States on their own. Some feared that the Shah's Iran would reach across the Persian Gulf and swallow them up, or that power struggles would undermine their union. But the emirates, under the leadership of Sheikh Zayed bin Sultan Al Nahyan, joined together as a single nation, the United Arab Emirates. The country's development model, similar to Singapore's, combines a measure of tolerance for Western freedoms with a firm and steady dose of authoritarianism. Powered by Abu Dhabi's oil and the commerce and construction in Dubai, the UAE has been a rare economic success in the Arab world.

MAKE YOUR WAY through Dubai's mammoth airport, its fluted white columns evoking an antiquity theme park, and step into the immaculate subway. There a vision of the future snaps into focus. The train pulls away into dark tunnels, but a few minutes later emerges into the desert light and jets through a city of gleaming high-rises, most of them erected in recent years. To the left, the Khalifa Tower, or "Burj," the tallest building in the world, stretches upward. It grows narrower with height, and turns from silver to golden, like a twisted blade of wheat, as the sun sets.

The story behind that tower captures the relationship between risk-loving, free-spending Dubai and its richer and more staid sibling, Abu Dhabi. It was in 2002 that the ruler of Dubai, Sheikh Mohammed, opened a legal door that led to a rampaging real estate boom. He gave foreigners the green light to buy land in Dubai. Billions poured in from the troubled lands in that tortured part of the world, from Pakistan and Syria and Iran and Iraq, and from Europe and America, too.

"Put it this way," says one European executive for an American technology company in the country. "Say you're sitting on a lot of money, and you're in Pakistan or Iran or Egypt. Where do you put it?" It's safer as an apartment building in Dubai, or a Lamborghini dealership—even as an indoor ski mountain in Mall of the Emirates.

Dubai was friendly to business. Its working language was English, the common tongue of its Asian working class, which makes up some 80 percent of the population. Under a special clause in Dubai's commercial code, many foreigners could create businesses there. What's more, Dubai was largely tax-free (though the government drew revenue from plenty of fees and other services).

The boldest expression of Dubai's success, and of its ambitions, was to be the Burj. It would soar over the city's skyline as a signature of the emirate, a stylish exclamation point. Designed by the American architect Adrian Smith, the Burj would climb higher than any structure erected by humans. (Again, for Dubai's brand, "first" and "biggest" have always been essential elements.) During the early 2000s, as the Burj grew ever higher, the pace of construction elsewhere throughout the city was frantic. Hundreds of thousands of workers, most of them from the Indian subcontinent, poured in for construction jobs. Many such workers were exploited by both recruiting agencies in their home countries and employers in Dubai. This tarnished the emirate's brand.

But that didn't slow the boom. Money was easy. As in other frothy markets, from Las Vegas to the south of Spain, developers borrowed ravenously, as did the government. The Burj was crowned in 2008, just as global markets crashed. Credit markets tightened around the world and lenders called in loans. Dubai Inc. found itself teetering on the brink of insolvency.

Sheikh Mohammed trekked to Abu Dhabi—a humiliating visit, no doubt—and asked for a bailout. He returned with a commitment for $20 billion from the Abu Dhabi ruler and president of the UAE federation, Khalifa bin Zayed Al Nahyan. But there was a price. The record-breaking skyscraper, that tribute to Dubai's soaring ambitions and possibilities, would no longer be called the Burj Dubai. Instead it would bear the name of the man who saved the debt-ridden emirate. The gleaming tower presiding over Dubai would be the Burj Khalifa.

Dubai recovered in a hurry, and the construction fever returned as if the bursting bubble had been nothing more than a bad dream. Looking to the future, Dubai considers Expo 2020 to be its coming-out party as a leading global city. Central to Dubai's rising status is its place as a trendsetter in mobility.

Being first in anything is a gamble, but Dubai is built on them. Most have paid off. In the 1970s, when Dubai was a fraction of its current size, Sheikh Mohammed's father, Rashid, didn't just double the emirate's port capacity. He quadrupled it, and then dredged an entirely new port for traffic that did not yet exist. That traffic soon arrived, turning Dubai into a vital hub between Asia and the West, and a gateway to the riches of Arabia.

In a similar gamble, Rashid's son, Sheikh Mohammed, bet big on a start-up airline in the late 1990s. Most airlines, he knew, operated in the red. National airlines, like Alitalia and KLM, seemed structurally incapable of making money. Yet Emirates grew into a global giant. Today it operates as a massive conveyer belt, delivering jumbo loads of travelers to Dubai. Some carry on to Africa or Asia. But a lot of them *stay*. In what is perhaps the most unlikely of Dubai's triumphs, this sweltering patch of desert largely devoid of history has grown into a tourist hub. It markets its sun and beach, its refrigerated shopping malls. Equally important, it offers access to the discreet use of alcohol and other pleasures hard to come by in neighboring countries like Saudi

Arabia and Iran. Even when summer temperatures outdoors climb toward 120 degrees Fahrenheit, the local air-conditioning is so powerful that malls and hotels are not only cool, but even breezy.

Growing numbers will flock to Dubai, Sheikh Mohammed is betting, if visitors start posting on Facebook and Instagram pictures of themselves being ferried around in autonomous pods fitted with coffee bars or nail salons, or flying in a winged air taxi past the Burj Khalifa on the way to the beach. The future's a winning brand. California rode it for the best part of a century. Why not Dubai?

IF YOU HAVE any doubts about Dubai's commitment to mobility, a visit to the Roads and Transport Authority should lay them to rest. The building, molded by curved sheets of glass, is grander than most government edifices in Washington. The domed entrances to the RTA look like something Han Solo would stride through on his way to Jedi headquarters.

Outside the executive offices upstairs, oddly enough, the RTA feels much like a run-of-the-mill bureaucratic enclave. A well-used coffee machine stands in the hall, next to a paper shredder and an overflowing recycling bin. The carpet is industrial gray. It's a reminder that with all the focus on mobility glitz, the RTA also manages the everyday grind of vehicle registration and drivers' licenses. Most Dubaians, after all, still get around in cars that they operate themselves. The soundtrack of Dubai, like that of most modern cities, is the roar of car engines.

This is one of the first points that Ahmed Bahrozyan makes. He's the chief executive of public transport at the RTA. A tall, thin man, he wears the traditional white robe, or *thawb*, down to his ankles, and a white headdress, or *ghutrah*, anchored by a corded black crown. He speaks colloquial American English, which he perfected in the early 1990s as a student at the University of Denver.

Unlike London, New York, or LA, he says, sitting down to tea in his expansive office, Dubai doesn't want to discourage driving. "We don't want to make it more painful for people, or to penalize them."

It makes sense. The world's big, established cities, after all, are struggling with congested highways and bridges, many of them built for the 1960s or even earlier. They're legacies, and they're not sustainable. Dubai's highways, by contrast, are new, and still beloved. They stretch like black ribbon through the desert. They're some of the best in the world—and they won't be reclaimed any time soon for green space or pedestrian walkways.

The memories of unpaved roads are still too fresh in Dubai. When Bahrozyan was a boy, and the city didn't extend far from the shores of the inlet, or Creek, that provides its historic harbor, the dominant form of public transit was the fleet of wooden boats, or *abras*, that ferried passengers and cargo back and forth. Growing numbers of Dubaians were buying cars, but they'd often get stuck. "We'd have to get out to push them through the sand," he recalls. Well into the 1970s, drivers between Dubai and Abu Dhabi braved an unpaved trail along the beach, occasionally swerving around camels. In wet weather, vehicles sometimes sank down to their axles.

So no, Dubai still wants its nice new roads—but with less congestion. In fact, much of the push toward autonomy is not to coax wealthy Dubaians out of their Porsches and Ferraris, but instead to make their drives more pleasant and efficient—to ensure "happiness," as Bahrozyan stresses. The key, at least in these early days, is to find alternative transport largely for the Tamils and Pashtuns, Bangladeshis and Filipinos; in short, to move a big slug of Dubai's foreign-born working class off the roads.

The first big step was the metro. It opened in 2009, with Dubai in full crash mode. Many back then saw the new train line as a brand-building exercise and a waste of billions. "A lot of people didn't think it would be used much," says Bahrozyan. But the spotless trains, which arrive like clockwork every two or three minutes, run full, and they help alleviate traffic on the fourteen lanes of Dubai's main artery, Sheikh Zayed Road.

One detail you might miss on your first ride is that Dubai's trains have no drivers. In fact, the metro constitutes a major piece of Dubai's autonomous strategy. This might seem a little like cheating. When HE Al Tayer told the crowd in Los Angeles that 25 percent of the transport in Dubai would be autonomous by 2030, many of us pictured thousands of autonomous cars plying the desert highways—not subways. Autonomous cars are a big part of the plan, as we'll see. But Dubai's autonomous future is taking off first in ways that most people find harmless, even banal.

For decades, hundreds of millions of travelers around the world have been riding in autonomous trains from one airport terminal to the next. Few give it a second thought. It's much like an elevator, another autonomous technology, but horizontal instead of vertical. Dubai's metro is simply an extension of the same technology—but with one notable distinction: it was listed in Guinness World Records in 2011 as the longest autonomous train in the world. (Dubai would have it no other way.)

Still, there's nothing titillating about the driverless metro, much less scary. This is a model for the path ahead. In many cases, the rise of this mobile age of robotics is likely to occur in a series of steps, each one unthreatening, often to the point of tedium. The next step for Dubai, says Bahrozyan, is likely to be autonomous car-sharing services. The first ones will likely travel along predetermined routes—basically virtual tracks. So they'll feel a bit like trolleys—more boring than scary.

This is a central challenge for tech adoption. While humans love our smart machines, many of us grow nervous when they get too smart. We don't want them nosing into our lives indiscreetly and blabbing, much less ordering us around (or for that matter, taking over the world). So for some at least, the word "robotic" is tinctured with existential angst. That fear, from a roboticist's point of view, is anything but market friendly.

Here's where the paradox arises: For robotics to carry out the endlessly complex task of driving a car, the software powering it must be extraordinarily powerful and nuanced. However, when it comes to introducing the technology to a skeptical public, it's preferable to hide the AI. Robots are less threatening if they act like dumb machines.

Bahrozyan predicts that these public relations concerns will fade away and that robots will soon be taking the wheel. "There's too much technology and investment to it for [autonomy] not to succeed," he says. He's already working with automakers around the world, from Tesla and Toyota to GM's Cruise, to launch pilot programs in Dubai. He expects most of these companies eventually to operate fleets of autonomous taxis. Step-by-step, they'll venture into the randomness of human mobility, mingling their machines with human drivers on streets where they must grapple with traffic, skirt the odd pedestrian, and master at least enough geography to navigate safely through Dubai.

The early customers? The metro crowd, mostly the working-class immigrants from different parts of Asia. It will cost less, he says, than they spend on cars.

Bahrozyan shrugs when considering the timetable. Even though government leaders, most notably the sheikh and HE Mattar Al Tayer, list target dates, they're based on (optimistic) guesses about the speed of advances in autonomous technologies, from batteries to AI. No one can fault any country for making sure robotic cars are safe before launching them willy-nilly. If they come a year or two later, Dubai will still race to adopt them first. The goal is to have at least pilot projects in place before Expo 2020. By then, the Hyperloop should have a ten-kilometer test track up and running. Naturally, it will perform near the expo and an even newer and grander airport south of the city. Full Hyperloop service, presumably, is to launch a few years later.

If the fleets of autonomous cars are not deemed ready in time for the expo, Dubai will no doubt hold plenty of showy demonstrations to highlight these coming attractions. At the very latest, the emirate will reach the future tied for first. It's a mandate from the ruler, one that is central to the brand.

With the exception of human-powered mobility, from walking to cycling, everything that moves in Dubai will eventually run on electricity. That's the plan. So Dubai, as you might expect by this point, is building nothing less than the world's largest concentrated solar project. It will feature the world's tallest solar tower. It's part of Dubai's strategy, according to Sheikh Mohammed, to be carbon-free by midcentury, and to become a global hub for green technologies. (Dubai's ruler, it must be said, does not lack for ambition, or chutzpah.)

The biggest concern, says Bahrozyan, isn't the electricity or the electric charging stations. No, he says, the biggest challenge is the battery itself.

An effective and profitable autonomous cab will be lugging people and things within the emirate for perhaps twelve to sixteen hours a day. That will require a massive consumption of power and a range of perhaps five hundred miles—all of them fully air-conditioned.

Batteries are an even bigger issue for Dubai's boldest mobility venture, or at least its most showy: drone taxis soaring across the desert sky. In recent years, Dubai has held much-trumpeted exhibits of the technology. Flights with a German-built drone, Volocopter, went without a wrinkle—and garnered lots of media footage and multitudinous hits on YouTube.

Despite this success, Volocopter still must compete with other powers in the industry, including Airbus, to win the first contract in Dubai. The emirate here has significant leverage. All such manufacturers are eager to lead the industry into the earliest markets. A contract in Dubai means a big leg up. It could spell early leadership in the industry, which could be crucial. So they'll compete, and Bahrozyan hopes eventually they'll offer competing services in the emirate.

The drones in their early years will serve largely as hood ornaments for cutting-edge mobility—and not yet a solution for traffic jams below. They might snatch the chosen few from the garden of the luxury Burj Al Arab hotel, which rises in a blue curve, like a giant wave on the Persian Gulf. From there, it might be five minutes to either airport, or a minute or two longer to the Emirates Golf Club. Long into the cooler desert night, guests might play its floodlit "Faldo" course—and then, presumably, hop on an air taxi to the beach.

Air drones, Bahrozyan says with regret, are the one technology not likely to make it in time for Expo 2020. Not that the Dubai hosts won't stage plenty of demonstrations. It's just a matter of time, after all.

WHILE WINNING THE mobility race as a consumer is the primary goal, Dubai has a higher ambition. Sheikh Mohammed says, and writes in his book, that he hopes to make Dubai a regional leader in artificial intelligence and autonomy, a Silicon Valley on the Persian Gulf.

This has long been the plan. The country has lured technology companies with a business-friendly and low-tax approach. A clear commercial code reduces fears and friction. Sheikh Mohammed has set up hubs with all kinds of incentives for denizens of the information economy. Internet City stretches about a mile south of the ski slope at Mall of the Emirates, and it has its own metro stop. Tech companies have poured in, including IBM, Google, and Microsoft.

Such tech giants are all too happy to use the emirate, with its impressive roster of direct flights, as a regional sales hub. But to Dubai's dismay, they don't count on it much for software development. That's far more likely to take place in nearby Israel, a global leader in mobility and all its component technologies. Israel benefits from a lively start-up culture and deep ties to the powerful Israeli military-industrial complex. For R&D in the Middle East, Israel is a no-brainer.

Building an information economy is harder and slower working than erecting skyscrapers and paving highways. The Emirates pour money into education, and lure foreign universities, including New York University and the Sorbonne, to set up local campuses. (For the universities, these are cash machines. They can charge rich Emiratis, Saudis, and Egyptians top dollar.) China's Alibaba is investing $600 million in the new Tech Town, which hopes to one day host scores of companies developing AI, robotics, and new mobility technology. But building a highly skilled tech workforce in Dubai will take a decade or two, in the best of cases.

For now, Dubai participates in the research by offering itself as an open and eager laboratory for cutting-edge technologies. This has its risks. But risks are Dubai's competitive advantage, and they always have been. Abu Dhabi's oil billions, much to that emirate's dismay, serve as collateral.

AT SOME POINT, when building the city of the future, skyscrapers can feel like old hat. So Dubai's rulers chose an entirely different form for their new Museum of the Future. The museum looms over the south side of Sheikh Zayed Road like a giant eye—a silvery oval eleven stories tall with a gaping hole in the middle. The new museum's walls appear etched with black Arabic calligraphy, which includes snippets from Sheikh Mohammed's declarations about the future. The sinuous Arabic letters double as windows.

The design comes from Killa Design, a ten-year-old architecture firm in Dubai. It's headed up by a South African, Shaun Killa, and the partners in the firm are a United Nations of architects. It stands to reason that Dubai would attract designers from all over. Architects there can work on some of the world's coolest projects, and they get built in a hurry, without a lot of meetings. What's better than that?

This is how Dubai draws talent. The architects work closely with Noah Raford, a PhD from MIT and chief futurologist at the Dubai Future Foundation. In a video on the foundation's site, Raford lays down an offer for global talent: If you have cutting-edge artificial intelligence, he says, or applications in virtual reality, and you want them featured in the museum, show us what you've got. That's the formula: talent, whether architects or scientists, convenes in Dubai for a well-funded chance to build the future.

The same model holds for mobility. Climb a little bluff from the museum and head into the ground floor of the swanky Emirates Towers. Just down the hall from a Starbucks is the home of the Dubai Future Accelerators—a large open space containing tech displays, work studios, and a small auditorium. A tech incubator there funds and houses technology start-ups for six-month sessions. More valuable than that, it gives them a chance to try out their mobility ideas in a real city.

On one side of the Future Accelerators, in one of the studios, sits an American named Brad Johnson. He's an affable software developer for Swim.AI, a Silicon Valley start-up that models the digital world of mobility. The goal, he says, is to move people and things in optimized flows, as much as possible like packets of data through the Internet.

The first step for building a predictive mobility model is to translate the moving molecules of a city into mathematics. That process is at the heart of the mobility revolution. It's how technologists increasingly will gain control over mobility and manage it. Yet the people who express molecules as math need to deploy their algorithms in real-world laboratories. Dubai supplies an entire emirate, a ready and willing test bed.

The standard model for networked mobility features a command center, where a team of data scientists processes all the movements in a city and attempts to optimize them. Such centralized command, from Los Angeles to Dubai, starts with traffic controls and public transit. Eventually it migrates to networked cars, scooters, and pedestrians.

However, Johnson says, when it comes to managing autonomous movement in a city, the centralized control runs into a big complication. Time. An autonomous car doesn't have a second or two for a faraway computer to crunch several billion data points and come up with an optimal pathway. The car needs to know immediately.

To face that challenge, Swim's goal is to distribute more of the intelligence to the millions of actors in the mobility drama—the cars, the subways, the traffic signals. In Swim's scheme, each one will carry its own statistical model of the world from its perspective. This will include data not only about its own patterns, but also about all the elements it comes into contact with. An autonomous car, for example, should "know" that pedestrians in Dubai regularly jaywalk across Al Fahidi Street, in the tiny historic district, at lunchtime. Each vehicle should be able to calculate, on the spot, what is likely to happen and how to respond. What's more, the autonomous cars and all the other intelligent assets throughout the city will share relevant information gathered from their sensors with their mobility brethren—with smart sidewalks, other cars, scooters, and most likely, everyone's phone.

This distribution of intelligence is called "edge computing." It's a growing field of AI and a key area of research at big tech companies from Google to Baidu. The goal is to provide the moving actors in the drama enough computing power, data, and cognitive smarts to make as many decisions as possible on their own.

Naturally, edge computing requires rivers of real-time data. This is one reason Brad Johnson is sitting at this laboratory. Dubai is swimming in data. It's central not just to political control, but to the entire economy.

Think of the emirate as a real estate developer and a landlord. Those are its central occupations, and it carries them out in a very dangerous neighborhood. Dubai sells itself as a refuge, a safe place, both physically and financially. Its business model values safety above practically everything else, and it stands to reason. If people don't feel safe in Dubai, the brand goes poof. Long ago, people used to vacation in beautiful Lebanon and bank there. Its reputation for safety is long gone, and who takes holidays in Beirut these days?

So Dubai maintains a high level of surveillance. Omnipresent CCTV cameras capture most of the movements throughout the city, down to those of the individual. At a tech trade show up the road from the Museum of the Future, Dubai's police show off surveillance cameras that read the faces in malls or the metro and match them against databases. This technology, widespread in China, even estimates a person's age while carrying out a sentiment analysis, calculating each individual's mood.

It's a laboratory for networked management: a tightly controlled population is wired into digital networks. The ruler and his team work out the regulations, with no delays or interference from the public. No hearings. No petition drives. No picketers. It's far easier to adopt new technology in Dubai, because the emirate runs more like a corporation than a country. The corporation's strategy is to do what it takes, and to spend what it must, to show the world how to move. In Dubai, new mobility is a strategic imperative.

8

Idiot Savants at the Wheel

It was a voice mail invitation from the robotics lab at Carnegie Mellon University, in Pittsburgh. "There's something you'll want to see."

This was back in 1997, when many Americans were sending their first emails from AOL accounts, and cell phones were an extravagance. Compared with what we have today, computing power was a joke. But it seemed prodigious back then, and CMU's robotics lab, headed by an irrepressible computer scientist named Red Whittaker, always had interesting projects cooking.

A few years earlier, a CMU team had sent an eight-legged robot named *Dante* to a volcano in Antarctica. Its assignment, once it reached the crest of Mount Erebus, was to climb down toward the lava lake in the middle and gather samples of the gases flowing from the volcano's vents. Later, a company connected to Whittaker's lab built a robot that rolled on its own through the poisonous wreckage of the Chernobyl nuclear disaster, taking radiation readings and sending back video.

Whittaker's outfit, like most of the leading computer labs in the country, received a healthy share of its funding from the Pentagon's research arm, the Defense Advanced Research Projects Agency (DARPA). It didn't take much imagination to see how autonomous mobility could help out in times of war. Roving robots could clear mines. Self-driving army trucks could ferry ordnance or supplies to distant outposts. If they got blown up en route, at least no one would get hurt.

The organizing theme at this turn-of-the-century stage of mobile robotics was danger. Rolling, walking, and climbing machines could handle some of our most hazardous missions, the ones that could get us killed. Equipping machines for dangerous work opened up a promising market.

But on this late summer day in Pittsburgh, a team of Whittaker's roboticists was veering in a different direction. They had rigged up an ungainly gray minivan with a long sloping snout, a 1990 Pontiac Trans Sport. It wore cameras like sidearms, and a laptop computer was propped next to the steering wheel. Dubbed *Navlab 5*, the minivan was ready, researchers said, to drive all by itself through the lush splendor of nearby Schenley Park.

Off we went. The human (non)driver, a grad student named Todd Jochem, sat with his hands off the wheel, his feet hovering near the pedals. He glanced regularly at the laptop to his right. The minivan navigated the park's sweeping roads with tiny jerks on the steering wheel, like a nervous driver after his second cappuccino. But it stayed in its lane. Its cameras followed the painted lines on the road, Jochem explained. It could even take cues from the darker pattern in the center of the roadway left behind by tailpipe exhaust and drips of oil. There were computers in the trunk that somehow interpreted all this information and told the car what to do.

Conceivably, this story could have ended up in an article at *BusinessWeek*, where we were both working at the time (one of us in Pittsburgh, the other in Rome). The key for a weekly magazine article was the so-called nut graph. That paragraph, usually following an anecdotal lede, was supposed to tell readers why they should keep reading, why the story mattered. Maybe it was due to a lack of imagination, but it wasn't clear at the time, at least to the uninitiated, why a driverless long-snouted minivan navigating Schenley Park made much difference to the world. For lack of a nut graph, the story of that robotic drive went untold—at least until now.

These days, roboticists from Red Whittaker's lab are spread throughout much of the tech world, from Alphabet's Waymo division to China's DiDi Chuxing. In 2015, Uber swooped into Pittsburgh and hired forty scientists and researchers from CMU's robotics center. Ford, a year and a half later, bought a majority stake in Argo AI, a Pittsburgh start-up with roots at CMU.

Autonomous driving, clearly, is no longer casting about for its nut graph. It's at the center of a pivotal industrial race. A study by Intel and Strategy Analytics estimates that robotic cars will become a $7 trillion industry by midcentury. A number like that, constructed on a series of assumptions about the decades to come, is little more than a wild guess. The important point it makes, though, is that this branch of robotics promises to be large and transformative.

The challenge is to encode work our brain carries out into software. For a long time, people pictured the brain as a lopsided ball, its different thought functions deployed like continents on a globe. It was drawn that way in old textbooks. We know now that the brain's operations are much more complex. Our thoughts and memories involve complex interrelationships among tens of billions of neurons, and the overwhelming majority of these interactions are yet to be deciphered. Nevertheless, that old globe with the different cognitive continents, nations, and principalities could serve as a conquest map for information technology. Like an imperial army, the technologists are conquering cognitive kingdoms.

They've been at it for centuries. The development of written language marked an early victory. If one of the continents on the globe represents memory storage, tracts of it were freed up for other work as ancient technologists developed clay tablets, papyrus, and eventually the printing press. When calculators displaced slide rules in the 1970s, digital technology conquered the nation of arithmetic. Later, with word processing, we humans retreated from the land of spelling (though some sticklers still keep in mind the handful of words that still trip up computers, such as the distinction between "there" and "their," and whether "its" should carry an apostrophe).

When road maps went digital, and especially when we started to carry them everywhere, on our phones and in our cars, technology laid claim to a vast continent of our cognitive terrain, the realm of navigation. It also introduced a paradox: while giving us the most precise data on our location, it eliminated our need to know where we were or how to get where we were headed. All we required was a destination. In that way, driving in the last decade or two has started to resemble flying. We step into a machine. We follow orders. We climb out and we've arrived.

Internet maps, unlike the Rand McNally road atlas or London's old A–Z maps, reflect not just the land but the network of activity on it. Highways turn red, signaling traffic jams. In these early days, the maps simply provide information. With autonomous driving, the map extends into the physical world, and yanks the wheel from us.

Returning to that brain on a globe, you could argue that our driving skills represent only a small country, a trifle compared with the continents of memory or numbers. After all, we've been behind the wheel for only a century, and most of us don't learn to drive until we're almost adults. (Many New Yorkers never get around to it.) Driving is an exotic skill, like riding a bike or playing video games.

Yet ceding control of our bodies to machines, and entrusting computers with our lives, is a momentous step. That acquiescence—or surrender—is much harder for some people than for others. Some of us will insist on waiting longer, to make extra sure, with 99.9999 percent certainty, that nothing goes wrong. A certain contingent will never be satisfied, and every widely publicized mishap will fortify their conviction.

So when does the part of society that says yes—including an eager new industry—fling open the door and loose the rolling robots on the rest of us? That issue is sure to fuel political drama in cities and countries around the world. Much of it will focus on comparative economics. Will areas that are slow to adopt fall behind in the global technology race?

The Chinese government is committed to leading in this arena of AI, among others. One way to lead, in addition to funding research, is to accept a slightly higher level of risk. If Chinese manufacturers and software companies can leap forward and turn their unrivaled market into the world's greatest laboratory for the technology, pressure to match them is sure to mount across Europe and North America.

By the standards of robotics, autonomous driving has largely been achieved, especially on highways. The robotic cars already circulating in Arizona, Dubai, Pittsburgh, and Guangzhou have driven millions of miles safely and prudently, while obsessively following the speed limits. Yes, the complexity in cities can still confound them, and there have been accidents. A mortal tragedy in spring of 2018, in the suburbs of Phoenix, gave the industry pause. Uber, whose autonomous Volvo killed a forty-nine-year-old woman, suspended its road tests, only to resume them six months later on a smaller scale and with more controls in place.

The challenge at this stage boils down to mastering the one-in-a-million exceptions: the reflective truck siding in Florida that can appear to be empty space, a hunk of ice falling from a roof in Berlin. Each society, meanwhile, will have to weigh costs and benefits, and decide how much risk it's willing to bear.

Some argue that the status quo itself carries lots of risk—more than one million around the world die in highway accidents every year, including some 30,000 in the United States. If automated cars could reduce that number by nine-tenths, to 100,000 globally and only 3,000 in America, wouldn't it be worth it? "That's the way an engineer would think," says Jeannette Wing, the director of the Data Science Institute at Columbia University.

The logic is impeccable. But numbers alone cannot carry the argument. If we unleash robots that kill people, albeit in fewer numbers than before, aren't we responsible for those deaths? After all, human error wouldn't be the issue anymore, but instead society's decision to sacrifice human life for efficiency or profit. When risks are involved, we often pretend, as a society, not to quantify them, because to calculate the odds seems tantamount to accepting them.

So what do we do? We plow ahead on autonomous driving. Eventually, we deem it as safe as can be, while acknowledging that every once in a great while things go wrong.

A YELLOW SIGN on a mountain highway shows an S-shaped curve. This is a primitive map, and hardly a faithful representation of the road. Instead it delivers a simple signal to the driver: "Get ready for turns."

Road cartography has evolved over centuries with a unifying purpose: to guide people from point A to point B. The mapmaker's customer is a fellow human being, who demands clarity and gets confused by too much detail. In making decisions about which information to highlight, and which to omit, the cartographer imagines the thought process of the traveler. Maps, like language, are symbols that bridge human minds.

But this is changing. The newest field of cartography is designed for a different user, a software program. Unlike a person, the navigation program for autonomous cars feasts on specifics—every squiggle, every raised curb, every passing lane, all of them detailed down to the centimeter. The map doesn't have to be pretty. Its data-obsessed user couldn't care less.

At the same time, maps for autonomous cars must be ever changing, capable of adapting to the conditions in the physical world. Keeping such maps up to the minute is crucial. Robotic drivers, as you might expect, are extremely detail oriented. They're also stunted, at least compared with us, when it comes to improvisation. So they need a constant stream of reports, about traffic jams, roadkill, and jackknifed tractor trailers. In that sense, they require maps that communicate four dimensions: the physical world—length, width, and height—plus time.

Perhaps the best way to understand the limitations of the robotic driver is to reflect, for a moment, on the multitasking genius of our species. Let's say a driver in a Prius comes across a fallen oak branch blocking her entire lane. She's talking on the phone, and she swears under her breath.

"What?" her friend asks.

"Oh, nothing, just a—" She cuts off the conversation for a second or two while putting her miraculous human brain—by far the most complex work of circuitry known in the universe—into problem-solving mode. It carries out a host of lightning-fast what-if calculations, running each one through its own risk-reward analysis. She quickly eyes the traffic coming the other way, scans the intersection ahead for cops, and calculates the Prius's turning radius. In an instant, she performs a rapid and illegal U-turn and hops onto an alternate route, then returns to her conversation. That's a daunting load of intelligence, and in multiple dimensions, from predictions to spatial analysis. No AI can come close.

In addition to our fabulous brains, we also have a primal relation to the physical realm we inhabit. We grew up on earth, and we learned the fundamentals of light, momentum, and gravity as toddlers. In many ways, we are one with the world around us. Our tools, including cars, can feel like extensions of our bodies. AI scientists struggle to simulate aspects of that intelligence with prodigious mountains of data and processing power. A computer "perceives" danger only as the output of a complex statistical calculation.

Returning to that street where the woman pulled the U-turn, let's now picture an autonomous vehicle. It's stopped in front of the same fallen branch. But the obstacle poses a much bigger problem for the AV. An illegal U-turn is out of the question. So is swerving into the opposing lane to go around the branch. "They have clear instructions to follow the traffic laws," says Wei Luo, the chief operating officer of DeepMap, a Palo Alto mapping start-up. That means that AVs have virtually no latitude for going rogue. (That's one big reason they'll be dramatically safer than us.) The sad result is that the AV might be stuck behind that branch for a while—and all because it didn't get the message to avoid that particular street. Its map wasn't up to date.

Luo has been developing the next generation of maps her entire career. A small, soft-spoken woman (with extraordinary patience for explaining technology), she grew up in the city of Xi'an, in central China. That's where farmers digging a well in 1974 came across the famous army of life-size terra cotta warriors, all of them buried two thousand years earlier to defend the emperor in his afterlife.

Luo got degrees in economics and urban planning in Beijing,

and then she crossed the Pacific to study at the University of California, Berkeley. She earned her PhD in geographic information systems, or GIS. This field is at the intersection of geography and data. It describes a place according to what's happening there. Historically, GIS was a sleepy outpost of geography, in part because amassing all the necessary data was a chore. In a seminal 1832 application, a French geographer named Charles Picquet illustrated with a color-coded map the penetration of a cholera epidemic in every district of Paris. It was a brilliant project, but laborious—and with each new case of cholera, it grew further out of date.

By the time Luo reached Berkeley, GIS was no longer on geography's fringe. It was 1999, and the Internet was promising endless new streams of data. People would be able to spot not only a place on a map, but also its status. Was it snowing? Was the diner open? Was the foliage in central Vermont at its peak? The list went on and on. Many of the applications were tied to business cases—and the eager venture industry, at the height of the dot-com boom, was pouring millions into them.

The dominant GIS was the digital road map. In 2006, Wei Luo went to work at Google. For nine years, she developed products and services at the lively intersections of geography and data. She started at Google Earth and later moved to Google Maps. There she worked on a project to match people's interests with their location, and to suggest where they might want to go. GIS had come a long way from Charles Picquet.

WE ALL KNOW those people who look at maps obsessively, checking and rechecking them, even when the Eiffel Tower or Yankee Stadium is looming right out the window. "OK, OK . . . Should be ahead on the right. . . ."

If you multiply their obsession by a factor of ten, or maybe a thousand, you can start to understand the connection of the autonomous vehicle to its map. It cannot move without it. As the AV makes its way across town, its map unrolls the world in fabulous detail, every parking place, every pothole. Along the route, the car matches what it sees with what it's expecting. Stop sign? Check. Bus stop? Check. As long as what it sees correlates with the map, the AV is on the right track. This is essential work, and not too complicated.

The mapping software also requires a reporting function, and this is where things can get messy. To keep the network up to date, each vehicle will constantly be on the lookout for changes. It will encounter them in an endless stream. "When you hold up a magnifying glass to the surface of the earth," Luo says, "it's always changing." Shadows come and go. Branches wave in the breeze. Road cones block an exit lane on Charing Cross Road. In each car, these observations of the changing world generate an entire petabyte of data per hour. That's enough to store 250 movies. But which of the changes matter? The mapping program must sift through this data and pick out consequential items. Such judgment requires a thick layer of AI.

It's not hard to see, even for a robot, that the eight-point buck splayed on the left eastbound lane of I-10 in San Antonio is something to report. The networked map can measure its impact: traffic backed up 1.35 miles from Seguin. This is worth noting. Other observations, however, are less clear, Luo says. How about the repaving of a sidewalk? Will that affect nearby traffic? Should the map reroute the AV? A tiny puddle at the side of the road doesn't matter. But how about one that's five times bigger? Or one hundred times bigger? What is the threshold for a meaningful puddle?

The mapping software has to answer that question, and thousands of others. Some of the answers are written as rules. But much of the intelligence is generated by machine learning. The software pores over thousands of miles of traveling data recorded by AVs, and it begins to correlate certain changes, like road cones and sidewalk sales, with perturbed traffic. Puddles start to make an impact as they grow, and the AI develops its statistical threshold. This measure oscillates in time, adjusting to minutely changing circumstances. (In that way, it behaves much like commodities indexes in financial markets.) As the AI processes more data, advancing from thousands into millions of miles, its grasp of what's important and what's not grows more refined.

But it must work fast. For cars moving at high speeds, every tenth of a second can be critical. A central question for network architects is how to deploy the intelligence. How much of the geo-data should the vehicle itself interpret, and what proportion should be uploaded to cloud-based AI? On the one hand, the cloud can harvest data from multiple sources, match them with historical patterns, and provide more intelligence. But even with the ultraspeedy 5G networks, which boast of transmitting data one hundred times faster than the 4G status quo, a back-and-forth between the cloud and a vehicle raises latency challenges. While couch potatoes can twiddle our thumbs for a stalled Netflix movie, autonomous cars have no such leeway. What's more, since network connections are never guaranteed, AVs must be equipped to interpret deviations from the base map for themselves and respond appropriately.

This new high-def mapping is only one facet of an operating system for autonomous driving. The entire package is called the full stack. In addition to mapping, it features other areas of cognition, each one its own massive field of research. One of the primary layers of AI interprets incoming data, making sense of the world a car is traversing. It coordinates with the car's map. Another plans the route, and yet another takes action. It drives the vehicle, with all the difficult split-second decisions that entails.

Specialists like DeepMap, or Carmera in New York, are zeroing in on just one level of the stack. They're betting that industry leaders, including the car companies, will assemble their full stacks from the best-of-breed players in each field. They face integrated operations, like Alphabet's Waymo, China's Pony.ai, Aurora, and a handful of others, who are building the entire stack. All these companies are richly funded, teeming with PhDs, and hurtling toward an immense but yet-unknown robotics market, one that is busy being born.

One problem for the entire industry, especially for venture-backed start-ups like DeepMap, involves timing. While such companies are making large investments now, the widespread use of vehicles that can drive in entire regions on their own (so-called level 4 vehicles) could be a decade away from meaningful scale. Level 5 vehicles, the ones that can go anywhere, are even further in the future. In the meantime, start-ups will be burning through their venture funding unless they find near-term markets.

The most logical solution is to shoehorn applications of their AI into today's cars. We're already seeing this, as cars offer automatic lane changing, parallel parking, and all sorts of alerts and alarms. This process will usher us toward autonomy, taking control from human drivers bit by bit. At the same time, vehicles will feature smarter maps, with more up-to-date road status information and smoother course corrections. These new maps, designed to communicate with software programs, will try to earn a living, at least for a few years, servicing humans.

This next generation of maps will anchor all kinds of new services. With augmented reality, maps could provide detailed information on the route, in both audio and video form, and related entertainment.

If mismanaged, though, some of the new offerings could become highly obnoxious. Overzealous maps might push for stops at commercial partners, whether at coffee shops or electric charging stations. A fitness app could point out ideal places, every hour or two, for an exercise break. Like certain apps on a phone, some vehicles might make such a stop automatically unless the driver adjusts the default setting. (Imagine being late for a meeting and realizing, with horror, that the car is exiting the highway to take you for a restorative stroll in a redwood grove or a Civil War battlefield.)

During this period of transition, keeping humans in the loop has an advantage. Technologists can get a read on people's preferences. And the maps themselves can learn from drivers' responses to their data. We humans might swerve around certain puddles, for example, and splash through others. In this way, those of us behind the wheel will be educating the navigation engines poised to replace us.

IN THE SUMMER of 2010, the basketball legend LeBron James left his native Ohio and signed a contract to play for the Miami Heat. "This fall," he famously said, "I'm going to take my talents to South Beach." He wasn't going alone. In fact, James and two other stars, Dwyane Wade and Chris Bosh, had assembled a dream team, one designed to win championships. It worked, and it created an enduring model in the National Basketball Association. While management pays the bills, basketball's superstars organize their own teams and attract supporting talent.

Much the same dynamic occurs in the booming markets for artificial intelligence, including autonomous cars. Superstars dominate the field. They brandish top credentials from Stanford, MIT, or Carnegie Mellon, and they head breakthrough projects at the big data companies, like Google's parent company, Alphabet, or China's search leader, Baidu. Like LeBron James, these stars can raise money, attract talent, and build a potential champion.

In established markets, the idea matters far more than a university or job credential. Traditional entrepreneurs are experts at pitching what they'll be selling, whether robotic vacuums or exotic flavors of frozen yogurt. They can discuss target markets, costs, and prices. But the AI to drive a car is not a thing to sell. It's an intelligence yet to exist, one that develops and manages a set of skills. What investors are betting on, in essence, is that one company's AI will turn out to be smarter than the others', and that it will be more efficient (or, to use LeBron James's terminology, more talented). The common strategy for creating this intelligence is to spend money on a superstar, or a few of them.

This brings us to the Guangzhou offices of one of China's big bets on autonomous software, Pony.ai. The company faithfully follows LeBron's dream team template. The superstar is the cofounder, James Peng. He oversaw autonomous driving as a chief architect at Baidu. Before that, he worked at Google for seven years. He got his PhD at Stanford, as well as an undergraduate degree from Tsinghua University. Peng's résumé checks every box, emphatically. He's the CEO.

Then there's the chief technology officer, Tiancheng Lou. He's a competitive coder who goes by the name ACRush, a two-time winner of Google's celebrated Code Jam. The Pony.ai team includes Andrew Yao, China's only winner of the Turing Award, the equivalent of the Nobel for computer sciences, as an adviser. He has a PhD in physics from Harvard.

As soon as Peng built this core team, investment capital flowed in. Pony.ai raised more than $200 million within eighteen months. In 2018, it reached a valuation of $1 billion before earning a single dollar or renminbi of revenue. The word for a billion-dollar start-up is "unicorn," and the entrance to Pony.ai's Guangzhou offices features a large statue of a brooding rhinoceros. It's not nearly as graceful as the mythical horselike variety but, strictly speaking, a unicorn all the same.

Initially, Pony.ai established twin headquarters on both sides of the Pacific, one in Fremont, California, just down the road from Tesla, the other in Beijing. Then in 2017, the southern city of Guangzhou opened up twelve square miles of urban territory, an area with a population of 400,000, as a laboratory for autonomous vehicles. Pony.ai promptly mounted an operation there, just a short ferry ride up the Pearl River from Hong Kong. Pony.ai's goal is to power fleets of autonomous cabs with its software, and to operate them in cities around the world. But the first market is China.

China's long been a follower in the auto economy. For decades, the country imported foreign investment and know-how, and it produced mostly hand-me-down designs from Japan, Europe, and the United States. (Maoism was hardly a springboard for industrial dominance.) That's why few drivers outside Asia have ever seen a Chinese-made car, much less driven one.

But this is likely to change, due to two factors. First, the Chinese government is vigorously pushing electric cars in its enormous domestic market. China alone could account for 59 percent of global sales of electric cars in 2020, according to a J.P. Morgan study, and for 55 percent of a much bigger market in 2025. Despite trade friction, especially with the United States, Chinese automakers are hoping to storm Western markets with low prices and high-quality electric vehicles. In preparation, Chinese automakers have been hiring away top auto designers from Audi and BMW, among others.

At the same time, the goal of the government's Made in China 2025 campaign is to nurture a global elite in every major technology, including the full stack of AI for automated vehicles. So China's auto industry will feature electric cars, massive funding for AI research, unrivaled mountains of data, and the biggest auto market in the world—the only big one in a growth mode. It's a powerful combination.

Fueling China's advance is a race for dominance among its own manufacturing cities. Unlike the United States, which has a single iconic car town, China features numerous car manufacturing hubs. "There's a half-dozen Detroits in China," says Michael Dunne, an auto industry consultant with offices in Hong Kong. "Shanghai's the biggest, but Guangzhou is fast growing in size and quality."

From Guangzhou's perspective, if Chinese companies such as Pony.ai can develop dominant software platforms and marry them with its local car manufacturers, this proud capital of China's south, the city formerly known as Canton, could lead the next stage of global mobility—perhaps besting Beijing and Shanghai. That's the plan. It's why Guangzhou has opened a big chunk of urban real estate as a petri dish for autonomous technology.

On a sunny afternoon in Pony.ai's Guangzhou offices, the open work spaces behind the statue of the rhino are crowded with millennials hunched over computers. It could be an AI lab anywhere in the world. Presiding over the office is Harry Hu, the young chief operating officer. Hu came to Pony.ai from a robotics start-up where he ran finances. But in Guangzhou, he's overseeing a research lab. Hu wears a smile and speaks fluent English, but very softly.

The technologists in the lab are studying data pouring in from Pony.ai's fleet of two hundred autonomous vehicles plying the streets of Guangzhou's Nansha District. With its banks, office parks, and a big Sheraton hotel, the Nansha streetscape could be Tampa or an orderly Dallas suburb. The cars circulating there are Lincoln hybrids imported from the United States and electric sedans manufactured by China's Guangzhou Automobile and BYD. The goal is to have fleets of these cars providing taxi services in major cities by the early 2020s.

If you look at the stack of software as a series of cognitive functions, you'll note that its first job is to perceive its surroundings. Developing machine perception represents an entire field of AI, including machine vision. In autonomous cars, perception works hand in glove with the map. Compared with our own, an AV's perceptions are miraculous. It can "see" a mile away in startling detail, and it can calculate distances by the centimeter. The trouble is, it struggles to make sense of all that data. It doesn't "know" anything, and it attempts to compensate by crunching numbers.

It starts off its education millions of years behind us. Picture one of our ancestors, a young girl, taking a morning stroll on the African savanna. She detects a crouching form in the distance, partly camouflaged by the yellow grass. She recognizes the dark shape as a lion. She estimates its distance from her, and she calculates the speed with which it can cover that distance. If these reckonings are on target and she manages the risk smartly, she improves her chance, ever so slightly, of surviving and having children. Those perceptions, handed down to us over millions of years, have become second nature. They're fundamental for driving.

To replicate those human perceptions, Pony.ai's cars in Guangzhou are bristling with sensors, including three sets of eyes. The most important are the two LiDar units mounted to the car's roof. They each send 160,000 laser pulses per second in every direction. By measuring the time it takes each pulse to bounce back, the LiDar captures the distance of the forms around the car. The cameras, for their part, detect the shape and color of each of those forms and feed the information into an AI that identifies them: the cars, road signs, trees, and pedestrians. Ideally, it can distinguish a rolling dump truck from a stationary Dumpster. Meanwhile, radar captures the speed of everything that's moving. "If you add them together," says Hu, "it's a pretty good description of the world."

It takes powerful and sophisticated computing—so-called sensor fusion technology—to blend all that incoming data and make sense of it. Within microseconds, this software produces a coherent vision of the ever-changing street scene.

However, recognizing what surrounds the car is only the first step. The system must also map out the car's route, moment by moment, and guide it through what amounts to an endless obstacle course. Every block or two, traffic lights turn yellow, drivers tailgate, an old man carries groceries across the road. The car must respond to each of these challenges in order to fulfill its mission of getting to its destination safely.

Inside each of the Pony.ai cars sit two human beings. One is at the wheel, ready to take over in an emergency. The other rides shotgun, studying the screen on the dashboard, to see what the car is sensing and how it interprets it.

The AV is on an educational journey. The goal of each voyage is to confuse it, because each puzzling scenario is a learning opportunity. Most of them involve the idiosyncratic behavior of human drivers—the biggest wild card for an autonomous vehicle. The AV might be stymied, for example, by a four-way stop where a timid human driver refuses, for one reason or another, to take his turn. In such a situation, another human driver will communicate, maybe with an impatient wave and by angrily mouthing the word "GO!" Or perhaps the frustrated driver will flash his headlights or even honk. But what can an AV do? For engineers, that's one more problem to solve.

Or maybe one of Pony.ai's cars is circling one of Guangzhou's rotaries and doesn't slow down for a merging motorcycle. Fortunately, the motorcyclist hits the brakes, and a sideswiping crash is averted. But the incident gives Pony.ai's human drivers a start. It's true that the AV, by law, doesn't have to slow down. It's the incumbent on the rotary, and it has the right of way. This rule is programmed. Still, a prudent human driver would brake just a tad, because you never know. The oncoming motorcyclist might not know the rules, or he might ignore them. This one incident sparks a discussion among the programmers: Should they add a touch of rotary caution? Perhaps a tad more for motorcycles?

Each of these puzzles represents a new skill for the AV to master. The skills fit into a large library of scenarios, gathered over thousands of miles in Guangzhou, smaller tests in Beijing and California, and thousands upon thousands of computer simulations. At last count, the Pony.ai team had the scenarios divided into about one hundred groups, each one with numerous subgroups. For each scenario, Pony.ai's scientists build and maintain a slew of algorithms. Sometimes, says Hu, engineers enter into nearly theological debates over the proper category for a single scenario.

These scenario algorithms represent a mere fraction of the digital brain of an autonomous car. If you were able to dissect such a thing, you'd find its intelligence divided into two buckets: one for rules, the other overflowing with statistics. The rules are the laws and regulations, like speed limits and rights of way, that you can find in government driving manuals. If you violate them, you might get a ticket or crash. These rules form the AV's universe of certainty. They are coded in, and they remain beyond doubt or appeal.

The other bucket holds data. It is not knowledge, but instead the raw material for learning. Processing the data, the computer calculates an endless series of probabilities. This is the AI's simulation of "thinking." If a garbage truck ahead is slowing down traffic, the AI might carry out a calculation to weigh the pros and cons of changing lanes. What is the probability that doing so will result in a faster trip? What chance is there that another driver will swerve into the same lane? The answers come back in a blizzard of numbers, and from those the AV makes its decision and acts.

The computer leans on statistics to compensate for its shortcomings in the physical world—its utter lack of animal instinct. All animals, from ants to elephants, including ourselves, have an innate feel for time and space, gravity and momentum. Like the cave girl who spots the crouching lion, we have phenomenal pattern recognition. But we also build on what we can see, and we hone certain behaviors into instinct. This reduces the processing load in our brain, as the Nobel-winning psychologist Daniel Kahneman has written. It speeds up our responses. When learning how to drive, we think about how far back we should stay from the car ahead of us (and on occasion argue about it with our parents). But over time, this analysis becomes instinct, which is why a person can drive skillfully through a daily commute and five minutes later barely remember it.

Matching such human skills is a lofty goal for a host of machines running analysis, every bit of information it consumes either a one or a zero. These machines are climbing onto a huge and complex continent of cognition.

In addition to building these navigational brains, companies like Pony.ai also have to make their way as businesses. This involves cutting deals with automakers, staking out new markets, and, for most of them, launching fleets.

Running fleets is crucial, not just for the revenue, but also for the learning. Each autonomous service car, whether it's delivering food in a section of Beijing or working as a cab in San Diego, is also a student. It's gathering data remorselessly and occasionally getting confused. When the AVs are on the road working, it's as if they've graduated from college and are moving through grad school. The more vehicles each AI team has in circulation, the more it can learn and improve.

Operating fleets, however, brings its own set of challenges. Some of them cannot be solved by software. Say it's Mardi Gras in New Orleans. Two revelers spill out of a bar near Bourbon Street, green daiquiris in hand, and summon an autonomous pod to return to the hotel. They step into the car and, of course, they find themselves alone and unsupervised. What happens next? Maybe one of them drops his po'boy on the floor and inadvertently steps on it. Maybe his friend spills her drink. Perhaps they darken the windows and tell the machine to drive around in circles while they have sex. That's just one messy scenario. Other passengers might climb into the AV with their shedding animals, if not rabid ones. The possibilities for mischief are endless, and all of them spell an unpleasant, and even disgusting, trip for the next customers in the car. This threatens the taxi's entire business model.

In a technologist's ideal world, the full stack running the pod would feature a self-grooming app, maybe with robotic arms that could swab seats and sprinkle disinfectant on carpeting. But that's a long way off. In the short term, the AV service will attempt to outsource these maintenance responsibilities to human beings, its customers. This is hardly new in the networked economy. Social networks like Facebook and WeChat have built empires on content created by their customers. Getting customers to clean dirty cars might be a bit more challenging.

One tool to limit the troubles is surveillance. The vehicles are sure to have cameras on, 24/7. So they'll know who left the smeared hamburger wrappers on the seat. Surveillance might enable companies to give unruly passengers bad ratings. Some might blackball certain people from the service or charge them more money per mile. But Harry Hu, looking at the positive, considers a rewards program. "If you follow the rules for ten rides, you get another for free. We now internally are testing different motivation mechanisms," he says (sounding for the first time vaguely Maoist).

IT'S A STEAMY day in the swamplands of southwest Florida. In a new development called Babcock Ranch, a group of elderly people is lined up along a curb. Many wear baseball hats; some shade their eyes with magazines. A few lean on canes. They're waiting for a test ride in an autonomous car.

Babcock Ranch is a test bed for the technology. It's the brainchild of a developer named Syd Kitson, a former lineman for the Green Bay Packers. After football, Kitson turned to the booming development business in Florida. But it didn't take him long to see that autocentric growth was paving the entire state. He saw it as unsustainable.

Amid this growing network of blacktop, one patch of green endured. A huge tract, the Babcock Ranch, occupied much of the crucial aquatic corridor between Lake Okeechobee and the Gulf Coast. For a century, the Babcock family ran mining and cattle operations on some of it. But most of it was left wild, a habitat for alligators, herons, and egrets, and a greenway for the water flowing through.

After working at it for an entire decade, Kitson in 2014 orchestrated a deal with the Babcock heirs and the state of Florida. He'd buy the land from the Babcocks and entrust 80 percent of it, the wild part, to the state, as a preserve. On the remaining acreage, he would build a town of the future.

But not just another town. He hammered out a deal with Florida Gas and Electric to build an enormous solar farm on his land. Babcock Ranch, he promised, would be sustainable, designed for the postcar economy. As Kitson saw it, he was putting together a model for Florida's future.

The first two hundred houses are up. Like countless other developments in Florida, they have two-car garages. Many of their new owners drive to offices in Fort Myers, or up the coast to Punta Gorda. The auto economy doesn't disappear just because Syd Kitson says so.

In fact, sustainable developments like Babcock Ranch face much the same challenge as Pony.ai and the rest of the autonomous industry. They have to grow during a decade still dominated by cars. So the idea, Kitson says, is not to deprive people of cars, but to wean them away. For this, the new Babcock Ranch features a walkable downtown, one with a general store, an ice cream parlor, and a growing network of bike paths.

But the centerpiece of the strategy, a fleet of autonomous vehicles, should roll up sometime in the 2020s. As residents get used to summoning rides, Kitson predicts, they'll get rid of their first cars. Eventually, as the fleets extend their range beyond the Babcock acreage and onto the highways and boulevards of southwest Florida, "they'll give up their second cars, too."

The key is to prepare the population for this autonomous future, which brings us back to the crowd of Floridians standing at the curb. Working with the Society of Automotive Engineers, Kitson is holding three AV demonstration days. So many people have come for rides that some of them give up and go home.

Those who stick around take a slow, two-mile tour in an

autonomous SUV operated by a Virginia start-up, Perrone Robotics. The trip, up and down a handful of culs-de-sac, is about as careful and deliberate as that distant 1997 spin in *Navlab 5* through Pittsburgh's Schenley Park. The SUV circulates with the tiresome precision of a car in a driver's ed video. Except for the small detail that the nondriver has his hands off the wheel, the circuit is a big yawn.

That's precisely the point. The key to selling people on the technology is to establish it as prudent to a fault, a stickler for rules. Jeff Brandes, a state senator from St. Petersburg and a booster of autonomous technology, puts it this way: "People are terrified for the first minute, interested for the next five, and then bored." That's the way autonomy will insinuate itself into our lives, as a service in geo-fenced zones, whether Babcock Ranch or the Nansha District of Guangzhou. We'll know that the technology is gaining trust when passengers climb in, get bored, and surrender to sleep.

9

Shanghai: Petri Dish on the Huangpu

For a generation of Chinese tech researchers, study in the United States has been a rite of passage. When it comes to mobility research, many of them can draw on firsthand knowledge of the challenges in America, whether it's convoluted traffic in Cambridge, gridlock on Highway 101 south of San Francisco, or waiting an eternity for buses in Atlanta or Austin.

Xi Zhang, a chief researcher at Jiao Tong University in Shanghai, has a different perspective. He studied at Michigan Technological University in the town of Houghton, population 7,700. It's in the state's Upper Peninsula, 550 miles north of Detroit, just across icy Lake Superior from Thunder Bay, Ontario. The contrast with his teeming home city of Shanghai, with its sultry climate and population of twenty-five million, could hardly be starker. Recalling the Upper Peninsula winters, he hugs his arms to his chest and chatters his teeth.

Zhang knows, of course, that cities in America and Europe are far more complex than towns like Houghton. But he is convinced, he says, that a certain order prevails in the West, and that the chaos in China's megacities, including Shanghai, is unique, at least in the industrialized world.

It's true that Shanghai's madness has its downside. Legions of motorbikes zip along the streets and sidewalks alike. Since their electric motors make no noise, they can send foot traffic flying. Pedestrians dash across divided highways, sometimes dragging children by the hand or a dog on a leash. Before crossing at green lights, many drivers in Shanghai reflexively touch the brakes, because people sometimes fly through red lights. (Drivers used to honk their horns to warn oncoming traffic before jumping red lights. But even those alerts are rare now, since authorities, in a step to battle noise pollution, have outlawed "unnecessary" honking in China's major cities. Acoustic cameras take two-second videos of honking cars and send fines to drivers whose honks are deemed unwarranted.)

Yet Zhang argues that the chaos in cities like Shanghai, for all the problems it creates, also provides China with a competitive edge in engineering the next stage of mobility. If Chinese AI can be trained to navigate safely in Shanghai, if Chinese autonomous vehicles can skirt the bikes and motorbikes and anticipate a jaywalker about to dart from one corner to another, "they'll work anywhere in the world," he says with a broad smile. Conversely, cars trained on the more orderly highways of Arizona or even the freshly plowed roads of Michigan's Upper Peninsula are likely to be stymied by the complexity of China.

This argument—that China's complexity can be an effective defense—harkens back centuries. Three hundred years ago, during the Qing dynasty, Western traders wedged into their tiny enclave in Canton (now known as Guangzhou) were limited by the inscrutability of Chinese languages, Mandarin and Cantonese alike. It was a crime back then for Chinese to teach their languages to foreigners. Like Xi Zhang, Qing emperors believed that as long as their vast market remained indecipherable to foreigners, it would remain safely under Chinese dominion.

Questions of international competitiveness come up again

and again in Shanghai. With a focus on the national economy, the Chinese perspective is fundamentally different from what we've encountered elsewhere. For Helsinki, Dubai, and Los Angeles, mastering mobility is no doubt a key to thriving in the coming decades. And naturally, these cities are eager to outperform their neighbors, whether Stockholm, Abu Dhabi, or San Francisco. But their competitive focus is urban. In China, by contrast, even as Shanghai and Beijing joust for supremacy, the new mobility push is national and strategic. The Chinese are intent on leading in both artificial intelligence and mobility. It's a race. "China has a chance to leapfrog the West," says Bill Russo, a former Chrysler executive and now CEO of Automobility, a strategy investment firm in Shanghai.

On a warm autumn morning in Shanghai, under blue wisps of smog, a large crowd has filed into a daylong mobility conference at a central hotel. Zhang is a featured speaker. When his turn comes, he strides back and forth across the stage, with the open jacket, waving arms, and loosened tie of a commodities trader. He shouts out bursts of Chinese, brushing his bangs from his face as he calls up another slide.

His research, at the university's Intelligent Network Innovation Center, focuses on the foot traffic in Shanghai. He and his team are processing petabytes of street video through a machine-learning engine. The goal is to focus automated intelligence on the various classes and phyla of the Shanghainese pedestrian.

One of Zhang's slides snaps to life. It looks like a vintage 8-bit video game, with crude silhouettes representing pedestrians. The AI zeroes in on different body parts, he says: the ankles, the knees, the arms. It attempts to predict the movement of each body and where it is heading.

As with many AI projects, the challenge is to encode patterns that we humans recognize with a glance, often unconsciously.

Consider a Shanghai driver making his way through Saturday-morning traffic along the genteel sycamore-lined streets of the Luwan District, the old French quarter north of the Huangpu River. Shoppers stream along the sidewalks, some window-shop, others line up at meat shops, where butchers with cleavers neatly reduce mahogany-skinned Peking ducks into mountains of meat. Spotting an elderly man standing at the corner, leaning on a cane, the driver's brain carries out a lightning-fast risk calculation. What are the chances that the old man will dash across the street? Negligible, the brain concludes, and the driver keeps his foot on the accelerator. At the next corner, he spots a teenage boy. He's looking at his phone as he steps toward the street. The driver's foot instantly moves off the gas and hovers over the brake.

Zhang's research proposes to educate computers about this type of risk analysis. It's not enough, he says, simply to analyze the movements of the human forms on the streets and sidewalks. The AI must also establish each person's gender and approximate age, because pedestrian behavior varies greatly. "Men take more chances," he says, "but fewer if they're with women."

If you consider all the comings and goings in Shanghai as a gargantuan tapestry of movement, Zhang's pedestrian study at Jiao Tong University represents only a few slender threads. Its aim is solely to predict human movement, not to orchestrate it. The far larger project in Shanghai is to manage and optimize much of the motion in the city, from the throngs piling into subways at People's Square to the rush hour snarls on Middle Ring Road. For this, authorities in Chinese cities, including Shanghai, have a strategic advantage over every other economy in the industrialized world: an unrivaled mother lode of data.

EARLY EVENING IN Shanghai: Along the south bank of the Huangpu, in an area called Lujiazui, visitors snap pictures of one another, some framed by the river, others by the nearby cluster of skyscrapers. The spiraling Shanghai Tower, the second-tallest building in the world, looms highest, but it's only a head taller than a couple of the others. A few blocks away rises the iconic Oriental Pearl Tower. With its round belly and smaller round head, it stands on long, angling legs like a giant insect. It was China's tallest building as late as 2007. Now it's barely clinging to a spot in the top ten.

All the people sharing this evening in Lujiazui, the crowds strolling along the river and the thousands emerging from the metro stop at the Oriental Pearl Tower, are emitting rivers of behavioral data. Scientists in China can study people going about their lives at a scale unprecedented in human history. The question is how this knowledge might be used to reengineer movement across this immense city.

Let's start with what the data scientists can see. Perched like black birds on streetlamps, fastened to trees, and embedded in walls, thousands of surveillance cameras record an endless stream of motion. Some two hundred million of these closed-circuit TV cameras operate in China. Shanghai has its share. The surveillance itself is hardly unusual. To battle crime and terrorism, many cities around the world are focusing electronic eyes on activity in streets and stores.

However, Chinese surveillance is far more powerful than most other forms. It is linked to world-class facial recognition technology. This enables China to track the movements of nearly every person who sets foot in the country. In 2017, authorities in the city of Guiyang cooperated with a BBC journalist, John Sudworth, to test the technology. After Sudworth was marked as a "suspect," the AI identified his face and location within seconds. It took police in this city of 4.3 million only seven minutes to track him down.

On this early evening, the Starbucks along Lujiazui's bustling river walk is packed, and nearly everyone in the sidewalk café is fiddling with a phone. This addiction is hardly unique to China. But the Chinese appear supreme in their devotion to their small screens (perhaps a year or two ahead of the rest of us).

An outsize share of their attention is focused on a social networking site, WeChat. It's run by Tencent, one of the country's three Internet titans. (The other two are the search engine Baidu and the e-commerce giant Alibaba.) In the advanced, urban half of the Chinese economy, nearly everyone uses WeChat, whether for messaging, networking, or even paying for groceries or dumplings from street vendors. (Only foreigners and visiting countryfolk seem to use paper money in Shanghai.)

From WeChat alone, it's possible to trace where a person goes, her conversations with each friend, what she buys, what she eats, how much time out of her metro commute she's bent over her phone playing Tencent's top video game, *Honor of Kings*. As we've seen, other governments, including Dubai's, harvest loads of behavioral data. But China has by far the biggest and richest data sets on earth.

For a US comparison, it would be as if Facebook, Google, Amazon, and the cell phone companies all provided their data, willy-nilly, to government authorities. Privacy advocates in the West might argue that this is indeed what happens. However, in Western democracies, governments still must fight for access to data, in political battles and in the courts. And when reports surface that they've been harvesting it secretly, such as in the cases exposed in 2013 by Edward Snowden, angry debates ensue. The West struggles to balance society's security needs with the individual's right to privacy. In China, by contrast, such public-private struggles are minimal. Data is viewed as a strategic asset, like oil or foreign reserves. It is put to use for what the authoritarian government deems the common good. "We have a government that can take action," says Zhu Hao, the director of the Networked Mobility Committee at the Shanghai Transportation Industry Association.

So when it comes to reshaping urban mobility, authorities in Shanghai can draw on formidable assets: the world's mightiest, richest data sets, armies of AI scientists working to coax insights from them, and a strong government to turn the resulting statistical conclusions into laws and decrees. It's a combination of power and speed that mayors like Eric Garcetti in Los Angeles can only dream of as they scrounge for transit funding, haggle with fractious neighborhood groups, and fend off legal challenges.

Still, for all of China's modern architecture and cutting-edge science, the country is ruled by the lumbering Communist Party. Mao Zedong's face is still on the money. Planning decrees sound like time capsules from the 1960s. Consider the Master Plan for Shanghai, 2017–2035, approved by the State Council in 2017. It declares, "Shanghai is one of the municipalities under direct administration of the Central Government of China. . . . Shanghai will be an excellent global city and a modern socialist international metropolis with world influence."

In much of the world, central planning seems like a throwback. It belly flopped in the second half of the twentieth century, most notably with the fall of the Berlin Wall and the collapse of the Soviet Union. It was the free and vibrant marketplaces of the West that gave birth to technology innovations, from cars to computers. Entrepreneurs like Henry Ford and Bill Gates didn't follow a government's development plans. They disrupted them.

Planners are widely ridiculed in the West, especially in the United States. Planners plan, but it's the builders who build, and they're much more likely to follow a profit than a plan. Financial factors can usually change a plan or kill it. Nevertheless, even though the Shanghai plan, drawn up by thirty-eight committees, might seem like a bureaucratic joke, history suggests we take it seriously.

On a corner of People's Square stands the Shanghai Urban Planning Exhibition Center. In most cities, a museum dedicated to urban planning would be tucked away on cheaper real estate. But this giant exhibition hall, a large glass cube capped by a white roof shade looking like a whirlybird, is clearly central to Shanghai's identity. It's just across from the Shanghai Museum and next door to city hall (guarded by soldiers standing at attention). Stretched across much of the exhibition hall's top floor is a spectacular scale model of modern Shanghai. Visitors climb raised walkways above it for a bird's-eye view. The skyscrapers blink with tiny lights.

Few of the visitors wander into the small nook next to the model, where a TV screen plays a ten-minute black-and-white video on a loop. It's a 1984 film laying out the master plan for Shanghai. Created by the Shanghai Urban Planning and Research Institute, the plan was approved by Communist Party authorities in 1986. It includes much of the same language as the 2017 master plan, including the goal of turning the city into a "modern socialist international metropolis."

The film begins with a stark depiction of what ails Shanghai in 1984. The city is too crowded. Downtown is chockablock with tiny storefront merchants and manufacturers. In the narrow streets, people get in each other's way. They're traveling at different speeds, some on feet, others on bikes, and a third group piling into local buses. The result is urban mayhem. (Paradoxically, if you look at it from the perspective of this book, Shanghai in 1984 appears well on its way to the future: wondrously free from the car monoculture.)

To be sure, the Shanghainese of the mid-1980s are familiar with cars. In the 1920s and 1930s, when the great powers still occupied their bastions, or "concessions" of the city, they imported their Ford Model As, their Citroëns and Bentleys. But following the Communist Revolution of 1949—or "liberation," as many in China still call it—Chairman Mao and his cadres suppressed personal ownership, at least outside their own exalted circles. They viewed cars as unaffordable for the masses. So most of the population walked, biked, and used public transit. Sure, the crowds cramming into buses looked a bit uncomfortable. But nothing that a few hundred more buses couldn't fix. . . .

By 1984, however, eight years after Mao's death, planners in Shanghai were finding the status quo problematic. Under Mao, according to the video, factories and garment shops ringed the ancient downtown. These rings expanded, with time, into a series of concentric circles. The result was a crowded and insalubrious center, where people lived on top of each other and had trouble moving. The city teemed with what the movie calls "shabby factories." Thousands of families ran tiny garment businesses in their apartments. All they needed was a sewing machine and a table. Each one generated shipments, both of raw materials and finished products. This added to congestion.

The film segues to a hectic scene at an old airport. It's so cramped that people are waiting in long lines outside. Some shove suitcases through windows. It looks like a severely undersize facility in a developing economy, which is exactly what it was.

The first order of business, the 1984 film's narrator says, pivoting to the future, is to expand Shanghai's urban footprint. The city is to stretch out dramatically, creating sixteen different regions, each with its own green spaces and commerce. For this, the planners call for a new subway, with key lines opening by the year 2000. The more dramatic change is to create a new auto economy, and to put millions of Shanghai's walkers and cyclists and bus commuters behind the wheel. All the far-flung districts in the new Shanghai are to be connected by a series of new urban highways. (In the film they're called "trunk roads.")

Seventy years earlier, at the dawn of the automobile age, the legendary New York City powerbroker Robert Moses followed a similar strategy. He would tear down congested slums and tenements and build gleaming new highways and bridges, turning a crowded city into a sprawling metropolitan region. Highways would connect the masses to forests and beaches. New suburbs would take root.

Moses's vision spread to cities around the world. But by the 1980s, its downside was all too clear. Los Angeles and Mexico City were choking in smog. New York's commuters inched across Robert Moses's bridges into Manhattan and then struggled to find parking. Yet even this late in the game, Shanghai's Communist planners were joining the rush. They saw the automobile, both for manufacturing and mobility, as central to Shanghai's future.

The point here isn't whether this was a good strategy or a dumb one. What's relevant, as we look forward, is that the Communist Party and its planners pulled it off. Today's Shanghai, this sprawling megacity, is pretty much what they drew up. It has the new districts, with their playgrounds and shopping streets, all of them joined by ten-lane trunk roads. Shanghai has graduated, if that's the appropriate term, to a full-fledged status as a car town.

At the same time, its metro, a mere proposal in the 1984 movie, has grown into the world's largest, with 397 miles of track. Trains usually arrive within a minute or two. The backward airport, where travelers once heaved their luggage through windows, is now a distant memory, having been replaced by two mammoth state-of-the-art facilities that make JFK or O'Hare seem like provincial airports.

China's central planners made good on their promises, while many similar plans in other countries languished: unfunded, unbuilt, bogged in legal battles, or simply shelved. Take, for example, a bold plan for the future Los Angeles, drawn up in 1970 under then-mayor Sam Yorty. Accompanied by drawings that look like sketches for a colony on Mars, the plan calls for an entire complex of gardens and parks on downtown rooftops. It envisions pedestrians strolling on webs of walkways above the traffic, and highways supplemented by an extensive rail system. Most of these plans remain stubbornly in the realm of what might have been.

Shanghai, by contrast, transforms before the world's eyes. Its plans, drawn up by committees and OK'd by the Communist government, seem to come to pass relentlessly, like the seasons. So as we look ahead to the next stage of mobility in Shanghai and other Chinese megacities, why should we doubt that they'll succeed once again?

ON DECEMBER 1, 2018, a forty-six-year-old woman was flying from Hong Kong to Mexico. She faced a twelve-hour layover in Vancouver, British Columbia. Interminable airport layovers are hardly unusual for the legions of field reps and machine-part vendors who crisscross the Pacific. But this particular business traveler, Meng Wanzhou, was not just another suit wedged into an economy-plus seat. She was the chief financial officer for Huawei, perhaps China's preeminent technology company. Huawei was the world leader in network equipment and the world's second-largest cell phone maker after Samsung. She was also daughter of Huawei's founder, Ren Zhengfei, a legend in China.

You would have thought Meng could have found a more direct route to Mexico, or even flown in a corporate jet.

The long layover, it turned out, was the least of her problems. Canadian authorities, following an extradition request from the US Justice Department, arrested Meng and threw her into jail. She faced charges in the United States of committing bank fraud to skirt international trade sanctions against Iran. She, her company, and the Chinese government all vigorously denied these charges. The United States, they argued, was intent on hobbling a juggernaut of China's tech industry.

Whatever the truth, the arrest of Meng Wanzhou exposed the heightening competition between China and the United States for technological dominance. Ground zero for this contest is in mobility. Huawei is at the center of it.

For the best part of twenty years, executives at Cisco, the US leader in networking gear, had been complaining bitterly about Huawei. They accused the company of stealing technology, and warned that Huawei's tight (but opaque) relations with the Chinese government represented a danger. In 2012, the US House of Representatives Permanent Select Committee on Intelligence declared Huawei and a second Chinese network company, ZTE, national security threats. The committee concluded that their equipment could be used by the Chinese government for spying on Americans. It's a charge the Chinese hotly deny.

Still, for years the spat among network equipment manufacturers seemed like a sideshow. In the rest of the digital economy, the United States and China both thrived, and without stepping too much on each other's toes. While American Internet giants like Google, Facebook, and Amazon were unhappy to be blocked out of the world's largest growth market, they carved out their spheres of dominance in much of the rest of the world. Their

Chinese counterparts, meanwhile, prospered behind China's Great Firewall, with Alibaba and Tencent maybe crashing into the top-ten list of the world's most valuable companies. China continued to manufacture hardware for everyone, and it became the biggest market for Apple's iPhones. Tech giants on both sides of the Pacific got rich.

But the next stage of the Internet, featuring the mobility revolution, threatens to upset that comfortable status quo. Here, the charges swirling around Huawei become central to the drama. As networked intelligence spreads to the physical world—to cars, bikes, traffic lights, and indeed entire cities—giants like Huawei become crucial players.

The gateway for this transition is 5G technology. Such networks are designed to shuttle information as much as one hundred times faster than the 4G systems they replace. Moving far beyond smartphones, 5G will eventually connect a chattering universe of machines, many of them on the move, along with growing constellations of sensors.

If you compare these networks to biological systems, as marketers and consultants invariably do, the electronic sensors serve as the fingers, noses, ears, and eyes. They register the honk of a horn, the image of a double-parked truck, the whiff of an electrical fire, and everything else that machines can count and measure. That information flies through 5G networks, which amount to the network's nervous system. Its principal destination is the command center, which manages all this incoming information. In terms of computer science, it runs what would be the operating system for a city. In life systems, it's the brain.

This is the digital biology of the city. Its components—sense organs, nerves, and brain—all represent mammoth global markets. Future mobility platforms will orchestrate the movements of our bodies and our things, and among the torrents of data they feed on will be our shopping habits and friendships. Mobility networks, after all, will not operate independently from the data worlds we already know. Instead, they'll sit on top of them, or even consume them.

Say a university student in Shanghai is pedaling through the city's Bund district, north of the river. He's on one of the legions of orange GoBikes in the city (many of them strewn and abandoned in weed patches or back alleys). Based on his previous movements, a command center will be able to predict that he's heading to his best friend's house, or maybe to a (not-so-secret) lover's. It might suggest that he pick up his favorite beer at a convenience store en route, and deliver him a digital coupon for it.

Every day, these networks will worm their way more deeply into our lives. Many of us, no doubt, will object to certain aspects: the intrusions, the surveillance, and the control. In the early days, especially, some of the navigation will seem clueless, like the stories of Google Maps steering people to the wrong Portland or headlong into rivers. Still, mobility networks will grow smarter, especially in data-rich cities like Shanghai. Many users will find the benefits convenient, even indispensable, much like those provided by smartphones today. And who rebels against *them*?

A who's who of tech companies is racing to outfit cities with these new digital body parts. Sales reps from IBM, Hewlett-Packard, Siemens, Samsung, and scores of others land every day in Dubai, Singapore, Vancouver, Vienna, and dozens of other mobility hot spots.

But Chinese companies are setting out in this global race with a big advantage: they have multitudes of cities on their home turf to wire. China's fifteen largest cities together account for more than 260 million people. Each one of them, whether giant Shanghai on the east coast or Sichuan's provincial capital, Chengdu, in the southwest, is a laboratory for networked mobility.

So are a host of entirely new cities, like Xiongan New Area, south of Beijing. They're being designed for new mobility. They'll have electric charging stations, bike paths, dedicated lanes for autonomous fleets, and surveillance cameras to capture every smidgen of mobility data.

The Chinese government is bankrolling this development. One of the goals is to turn companies like Huawei into global champions. By 2019, Huawei had contracts for 5G networks in sixty-four countries. A further step would be for other Chinese tech companies, including its Internet leaders, to crunch mobility data in cities around the world and optimize navigation.

Consider the strategic stakes of the tech race ahead: the companies that run mobility command centers will be managing traffic signals, fleets of autonomous vehicles, trains, ambulances, air taxis, the deployment of police, the flows of pedestrians, the manhunts for suspected criminals. In a very real sense, they will be on top of everything that moves in a city, controlling much of it. Considering that the world's population, increasingly, is moving to urban areas, these tech companies will be ordering much of human life.

If trade battles worsen in coming years, countries could find themselves with adversaries running their ports, industrial supply chains, and markets, not to mention their traffic. If you take this to a dystopian level—which risk analysts do as a matter of course—you can conceive of an adversarial company tweaking its AI algorithms to jangle a rival country's mobility network, creating traffic jams and wreaking industrial havoc. Even if cities are spared such nightmare scenarios, the strategic issues weigh heavily. A lot of power is involved.

It is in this competitive context that the US government took steps against Huawei. Under government pressure, Google pulled its Android operating system from Huawei phones. And in 2019, as Meng's lawyers in Canada challenged extradition, the Trump administration pressed Western allies to shun Huawei, citing security concerns.

American strategic concerns, at least in theory, should have had nothing to do with Meng's arrest. Nonetheless, it's certainly easy to understand why the charges against Meng sparked deep suspicion and a surge of angry patriotism in China. On the social network Weibo, according to reporting in Quartz, one commenter quoted a Chinese blockbuster movie, *Wolf Warrior 2*: "Chinese citizens: When you are in danger overseas, don't give up. Remember, there is a strong motherland behind you."

IN THE LATE 1980s, a young PhD from Shanghai named Joseph Xie landed a job at Intel, the semiconductor powerhouse, in Santa Clara, California. Xie worked on teams designing the x86 chips, the silicon foundation for an entire age of personal computing.

He returned to Shanghai the following decade and plunged into the domestic chip industry. If China was to make the big jump in technology, eventually matching and surpassing the United States, it would need an Intel or two of its own. The digital economy—after all these years—was still built on silicon.

In 2012, when Xie was already in his fifties, he and his partners launched a chipmaker, Shanghai Quality Sensor Technology Corporation (QST). The company would focus on microelectromechanical systems, or MEMS. As the name suggests, MEMS are tiny machines, etched and inlaid on silicon at nanoscale.

To understand the outsize role MEMS play in modern life, imagine trying to piece together all the machinery inside a smartphone: the camera with its sophisticated light sensors, the microphone, the motion sensors and gyroscopes . . . the list expands every year. All these tiny machines are embedded on MEMS. They miniaturize not only the size of this gadgetry, but also its price.

When Xie and his partners founded QST, they could see the coming revolution in mobility, and they envisioned how it would drive the global market for tiny machines. In the coming decades, trillions of sensors would be planted into nearly everything, from sidewalks to dockless bikes. Autonomous vehicles would be chockablock with them. The entire world and most of its human population would be reporting on its minute-to-minute activity. MEMS would be ubiquitous.

On a drizzly late-autumn night in Shanghai, with a trade war already brewing between the United States and China, Xie joined us for a roundtable discussion about the future of mobility in his native city. He arrived wearing a broad smile and a brush cut that's turned gray. He spoke fluent English, with an American lilt. His company, he said, was still small, on a global scale, far behind industry leaders like Texas Instruments and Hewlett-Packard in the United States, and STMicroelectronics and Bosch in Europe. But it was primed for growth, especially in Shanghai.

During this reporting trip to China, we were still debating which Chinese city to focus on. Tech experts steered us toward Shenzhen in the far south, right next to Hong Kong. In three decades, it's grown from a fishing village into a megapolis, the heart of China's tech industry—and with the largest fleet of electric buses in the world. Tencent is headquartered there. If we wanted to see the China of tomorrow, Shenzhen was a natural. Or we could take it one step further, and explore one of the brand-new cities under construction. That might point us to Xiongan New Area or Chengdu Great City, a new suburb in Sichuan.

But Xie argued for Shanghai. The cities under construction, he said, were like blank pieces of paper, similar to Dubai. "Everything's new there," he said. "Even the people are new." He maintained that the more fitting target for a mobility revolution was an established city like Shanghai, with its huge industrial base. Shanghai produces more cars than Detroit. It has a large state-owned automaker, SAIC Motor, which operates joint ventures with Volkswagen and General Motors; the company also produces lots of its own brands. The richly financed start-up Nio is out to battle Tesla in the luxury market for electric cars. A joint venture in the city has licensed Divergent 3D technology to print new lines of cars. Shanghai is a big place to change. And revolution, at its root, means to turn from one thing to another, not to start from scratch.

Shanghai carries a long history and its fair share of scars to the next stage of mobility. For a century, from the aftermath of the Opium Wars until the end of World War II, foreign powers, including France, Britain, and the United States, operated chunks of Shanghai as "concessions." Each one had its own system of transit. To this day, despite the steady work of bulldozers in recent decades, and the dizzying growth, the borders of those enclaves still affect the bus routes and layout of city streets.

In 2018, authorities in Shanghai opened a small section of roadway, barely 5.8 kilometers, to test automated vehicles and connected cars. That's small, but the city is hurrying to open up 100 square kilometers, an area nearly the size of San Francisco, in which to operate a smart mobility network.

The command center, naturally, is the crucial element for a connected city. Yu Lin, the vice chairman of an auto parts manufacturer, Shanghai SH Intelligent Automotive Technology, says that the eventual platform in Shanghai will tie together the bicycles, connected cars, buses, subways, and even pedestrians, to management movement in the city. He calls it a "one-stop smart mobility service."

As we walk outside from our roundtable, into the tail end of the evening rush hour in Wujiaochang, the new business hub north of the river, very little of Shanghai's efficiency drive is in evidence. In these early days, bumper-to-bumper traffic still knots the streets. The air smells of diesel fumes. The cars, following the law, refrain from honking. But their muzzled silence makes the beeps of swarming motorbikes stand out.

A two-hour drive to the west, in Alibaba's hometown of Hangzhou, the progress toward networked mobility is much further along. The e-commerce giant has developed an entire arsenal of AI applications, under the brand of "City Brain," to manage the movement of Hangzhou's 9.5 million people. According to Alibaba, the system can predict traffic and pedestrian flows an hour in advance at 90 percent accuracy. The first-order use of this data has been to open up roads for emergency vehicles, whether ambulance or police. Now such traffic reaches its destination in half the time as before, according to the city. The City Brain also tweaks traffic signals to speed up general traffic, with gains of 15 percent.

Alibaba's City Brain also has features sure to terrify privacy lovers. It claims to be able to identify more than 96 percent of pedestrians' faces within seconds. In a single minute, the system can race through sixteen hours of video, looking for faces, accidents, or crimes.

The commercial battle for command center contracts, the brainwork, is likely to be the ultimate competition in this stage of computing, just as it was in previous waves. The kings of the first PC age, starting in the 1990s, were Microsoft and Intel. The chip was the brain of the computer, and the operating system was what it knew. These companies dominated the industry.

When smartphones were gearing up, a decade later, the battle again centered on who would operate these machines—which company would provide and operate the brain. Early on, the betting favorites were Microsoft and Nokia, the software and handset champions respectively. But in the end, Apple and Google landed the market, in large part because they saw the smartphone market as something entirely new, not an extension of legacies from a previous age, whether Windows software or a Nokia clamshell phone.

The question is whether China can nurture global giants and dominate the next stage of the information age. Huawei and the three Internet giants—Baidu, Alibaba, and Tencent—are the most powerful. They're all spending richly on AI. Huawei alone poured $10 billion into such research in 2017. That's the equal of America's Alphabet, the parent company of Google and Waymo.

At tech conferences in China, from Beijing to Shenzhen, the constant question is whether a top-down system that nurtures champions, like China's, can prevail. The best counterargument, at least to date, is found in the United States, where companies bubble up from below and topple titans. The brutal nature of the free-market economy allowed the space for old champions like RCA and Zenith and Westinghouse to exit the scene, either in bankruptcy or consumed, piece by piece, by their growing rivals, who in turn transformed themselves into today's powerhouses. Most of America's current tech leaders emerged in the last fifty years, and many of them in this century.

One crucial difference this time around is that the mobility revolution takes place in the physical world, in cities like Shanghai. This is the domain of urban planning and, in China, five-year plans. They've worked before.

10

Squadrons of Pack Drones

On a December evening in 2013, millions of Americans saw a familiar grinning face pop up on their TVs. It was Jeff Bezos, Amazon's founder and chairman, being interviewed by Charlie Rose on CBS's *60 Minutes*. Bezos told Rose he had a surprise for him. "Let me show you something," he said. Then he led Rose and the camera crew through a door.

Sitting on a table was an Amazon-branded black drone. It had four legs and a yellow Amazon box hitched to its belly.

Rose lifted his hands to his face. "Oh, man . . . Oh, my God!" he said.

Bezos went on to introduce the so-called octocopter. There was no reason, he said, that such a drone couldn't be used as a delivery vehicle. Octocopters, he speculated, might be on the job within four or five years.

With this drone gambit, Bezos reinforced the branding of Amazon as a daring, trailblazing company. The e-commerce giant was poised to revolutionize not only global retail, but also delivery—that is, how we as a civilization move our stuff.

That all may be true, but those five years are long past, and we're not seeing too many buzzing octocopters depositing yellow boxes on front walks. Maybe someday they will. But delivery drones face a thicket of regulatory questions, not to mention power lines, kids with rocks, and maybe hunters with rifles. For Amazon, it's largely experimentation. Maybe someday, offspring of the octocopters will create a market we haven't thought of yet. That's how Jeff Bezos's company rolls.

Yet there's a fundamental difference between a drone delivering a package and that drone's first cousin, an eVTOL, carrying a commuter across Dubai or Manhattan. While the Amazon customer is no doubt happy to receive a coffeepot or garden sprinkler twenty-five minutes after clicking on it, he does not *lose* hours a week while waiting for his stuff. Delivery drones promise speed, but they do not deliver the precious gift of time.

If we look at it from a policy perspective, a central challenge in cities is less to speed up the movement of all our things than to untangle it. We each consume literal tons of stuff—groceries and clothing, fuel, office supplies, medicine, beer, furniture, and thousands of other goods. If a technological advance like drones makes it easier to move our things faster and cheaper, there's a risk we'll use such services extravagantly—summoning drones for a cappuccino or a pair of chopsticks.

That's our nature. When something is abundant and cheap, we splurge. Consider texts. While it seems like a distant memory, sending texts used to cost money. Now that they're free and limitless in most markets, messaging for many of us has become like breathing. You could make the case that the blizzard of messages disrupts our thinking. But at least it doesn't intrude on the physical world.

Moving stuff, by contrast, butts into our space in a big way—and far more so in crowded cities. It's one of the few inefficiencies of dense settlements. Throughout this book, we've seen how cities like Los Angeles and Helsinki are pressing to fill in empty areas and to add density. Concentrating more people in a small area enhances mobility and adds all kinds of savings. But on crowded streets, hulking delivery trucks and double-parked vans gum everything up.

The challenge for moving our things may well have less to do with building miraculous new machines and instead devising smarter processes. The trick will be to move less with more: to piggyback more of our stuff onto fewer and smaller vehicles, and to deliver maximum tonnage with the greatest efficiency. And as we'll see in the case of Indonesia, much of the delivery fleet may already be in place.

IN THE MEGACITIES of the world, whether Lagos or Mumbai, the privileged few are ferried across town in company cars with drivers. Some rate as limousines. The traffic's still miserable, but at least the wealthy can take their mind off it while someone else drives. This is true as well in Jakarta. And Nadiem Makarim, in 2010, was among the elite. A graduate of Brown University, Makarim was clearly on the rise. He was twenty-five years old and working for McKinsey & Company. Naturally, he had a car and a driver.

He hardly ever used them, though, because of Jakarta's paralyzing traffic. A World Bank study at the time predicted "total gridlock" in the city by 2016 if growth continued unabated. From Makarim's perspective, gridlock was happening already. He could either spend hours in the air-conditioned comfort of his company car, or venture out into the sweltering heat of Jakarta and *move*. The way to do this was to hail one of the ubiquitous motorcycle cabbies, the so-called ojeks. They zipped around and between the stalled lanes of traffic, usually reaching their destination in a fraction of the time as cars.

Makarim, as he described it later, studied these ojeks with the eye of an entrepreneur. In a near-gridlocked city, they offered mobility. That was worth a lot. So how could these two-wheeled taxis do more? He began to hang out with the ojek drivers, drinking tea and smoking cigarettes. They talked. It turned out that the ojeks' system, for all its wondrous weaving through Jakarta, was wracked by inefficiencies. It was unregulated, which on the one hand was a strength. Everyone could work, without applying for licenses or buying a medallion. But the lack of rules and regulations was also a weakness. The service was viewed as risky. Most women didn't trust it. That constrained the market.

A bigger problem was information. The drivers never knew where or when their next fare was. So from one ride to the next, they spent way too much time waiting. Many of them put in fourteen-hour days just to scrape by. In fact, it was thanks to this inefficiency that they had the spare time to drink tea with this American-educated consultant, and talk.

As Makarim saw it, each idle minute held potential. These drivers could be profitably engaged for far more hours of the day. The trick was to put them in touch with opportunities—jobs in which something or someone had to be moved. They just had to know.

In 2010, he launched Go-Jek. It started out as a small call center that connected riders with twenty drivers. As the business puttered along, Makarim returned to America, where he got a master's in business administration at Harvard. It was during that period that the global ride-sharing business took off. "Uber" became a verb.

By the time Makarim returned to Jakarta, venture investors were eager to fund Go-Jek and help turn it into something much bigger. The first cash infusion landed in 2014.

Makarim and his growing team soon shuttered the call center and turned the business into a smartphone app, like Uber. Business skyrocketed. In the following years, as funding piled into the company, new drivers signed up, first hundreds, then thousands. Makarim was building an entire mobility ecosystem.

He set about expanding the business. This made sense. He had a large and growing army of drivers and also an app on millions of phones, each one with a billing relationship. Funders, including Google and China's Tencent, were pounding on his door. With these assets—mobility, paying customers, and eager venture funding—what other services could he sell?

All kinds, as it turned out. Go-Jek branched into banking and all sorts of door-to-door services, including massages, toilet cleaning, and manicures. But one of the biggest opportunities—and the reason we're discussing Makarim in this chapter—was to move people's stuff. Toward the end of the decade, Go-Jek, now worth billions of dollars, was operating the largest food delivery service in the world. And its package delivery service, Go-Box, was growing into a major force. (A big test will come in other Asian markets, as it faces off against its Singapore-based rival Grab.)

Makarim doesn't merely run a mobility service. Much like Jeff Bezos at Amazon, he's busy building a logistics giant on a foundation of software, one that's turbocharged by machine learning. It handles thirty-five orders per second, and it has more than one million drivers. For each order, the system picks the driver and the route, calculating the optimal combination whether the cargo is a passenger, a package, or an order of lemongrass soup. It factors in not only speed and profits but also drivers' satisfaction. It continuously adjusts prices to demand, and it shows drivers where there's more money to be made. "If you don't have dynamic pricing, you have inefficiencies in your market," Willem Pienaar, a data scientist at Go-Jek, explained at a Google symposium.

The heart of logistics is resource allocation. But since Go-Jek's resources involve human beings, this requires study. Every hour of every day, the system is testing different methods to prod drivers and passengers alike, to keep everyone engaged and productive. Companies like Google, Amazon, and Netflix carry out similar analysis, whether to optimize clicks or page views or to entice viewers to review a movie. Such data networks are running today's biggest laboratories of human behavior.

From a city's perspective, an efficient network like Go-Jek alleviates the population's mobility pain. Frustrated residents have services that work. Nothing wrong with that. But one large and aggravating problem remains: companies tend to keep their precious mobility data to themselves. It's their most valuable asset, providing them with competitive advantages over smaller rivals. Sometimes they share morsels of it, but only reluctantly. Seleta Reynolds, the general manager of the Los Angeles Department of Transportation, complains of the same issue with ride-share companies like Uber and Lyft. "We should have been asking ourselves why these companies are loath to give us data," she says. "They're building businesses on our infrastructure, and we get nothing in return."

DATA IS CRUCIAL for moving things. The model for the next generation of urban shipping, the industrial supply chain, runs on bountiful streams of it. Masters of the supply chain, like Toyota, use these flows of information to choreograph thousands of suppliers as they send rolled steel, brake pads, barrels of paint, side-view mirrors. Some of these products arrive on container ships from America or China, while others are flown in from Europe. Ideally, each component shows up just when it's needed, because running inventory costs money. As supply chain software manages these flows, it ceaselessly hunts down inefficiencies and tweaks schedules, routes, or tonnages to reduce them. That's its core mission.

Such miraculous feats of logistics work because each manufacturer trades relevant data with every other company in its vast network. They provide sufficient guarantees, both technological and legal, to establish trust—and to pool their relevant information.

This is precisely what cities need—and distributed ledger-sharing technologies, such as blockchain, provide them with a potent tool. The blockchain was devised only in 2008, as a secure ledger for what would soon become the leading cryptocurrency, Bitcoin. As the name describes, the system features chains of digital blocks, each one carrying time-stamped transaction data. Crucially, each block also holds the encrypted identity of its two neighbors. This makes it virtually tamperproof, because an intruder cannot alter the data in one block without getting the OK from the others.

A blockchain generates trust. Since it can share data without divulging secrets like identities or bank account numbers, the possibilities are nearly limitless. In agriculture, for example, a blockchain can document the entire path of an avocado or a head of garlic, from the hour it was picked, whether in Oaxaca, Mexico, or Gilroy, California, until the second it rolls up on the checkout line at Safeway. This accountability should be enormously helpful for zeroing in quickly on sources of *E. coli*, salmonella, or terrorist tampering.

It's also fundamental for transit. Connected vehicles can trade information with each other about where they've been and where they're headed. In a centralized system like Singapore's, authorities will be able to coordinate shipments, directing each one along the most efficient route and schedule. This will grow far more powerful as machines take over the driving. Indeed, the biggest obstacles to traffic optimization are the capricious and undisciplined human beings at the wheels. (We tend to take commands from machines as mere suggestions, when we bother listening to them at all.)

With blockchain, many argue, a network can add efficiency without the need for a central authority. Vehicles will increasingly be able to communicate with each other. "Ideally, everyone operates on the fly," says Chris Ballinger, the CEO of Mobility Open Blockchain Initiative (MOBI), an industry-backed nonprofit. "It's open to all."

Logistical engineers can grow very excited when describing the efficiencies of the robotic future. In their animations, vehicles flow like a school of fish, merging seamlessly at intersections. In these visions, clunky traffic lights fade into history. The additional trust and communication present in this future have the effect of expanding capacity to the transportation infrastructure, but without the necessity of pouring a single pound of cement. Karl Wunderlich, a director at Noblis, a research nonprofit in Virginia, says that a neatly networked transportation infrastructure would quadruple the capacity of our roadways.

This network of movement in the physical world, as mobility mavens routinely point out, is analogous to the digital packets zipping through our information networks. In the digital realm, everything that's sent, whether it's an email, a video of the Super Bowl, or pornography, is broken down into billions of tiny packets, which are rocketed through the system and then reformulated. From the network's perspective, all these ones and zeros boil down to the same thing: content. The system is engineered to move petabytes of this content quickly.

Something similar is possible in mobility. China's Alibaba, for example, is building an AI system to manage mobility in Xiongan

New Area, the brand-new city south of Beijing. For shipping, the idea is to share capacity, with every company's parcels blended with all the others, much like packets in the Internet. Out they go from distribution centers, in small electric vehicles, each one ideally close to capacity. Some of them might be shipped where we least expect it. At night, for example, when subway systems are closed, could that capacity be used for crosstown shipments?

The biggest challenge for delivery is getting stuff to people's doors—the so-called last mile. Here humans are part of the equation. After all, we have brains that can respond to incentives. We also have legs. Most of us can move and carry things. Our own muscles no doubt will prove valuable.

Let's assume the Alibaba-powered delivery network calculates how much time and energy it can save by dropping stuff at a secure depot four blocks from a person's fifth-floor apartment in Xiongan New Area. How much incentive should it offer him to walk over for it? At first, it might establish a uniform rate. After a few million deliveries, though, it should have a much clearer picture of how far each one of us is willing to walk and carry—and how much to pay us for it.

ON A NOVEMBER evening in 2016, the sixty-seven city council members of Tampere gathered to consider a mobility project: a new tram. For more than a century, variations of this project had been debated in Finland's second-largest city. The first proposal surfaced in 1907, and many in Tampere expected trams to be circulating within a decade. But World War I broke out, and the tram was postponed. Through the rest of the twentieth century, history continued to rear up, putting one tram project after another on ice.

By 2016, though, it appeared the tram was finally ready. It was just a question of getting it through the city council. A couple of

Finnish filmmakers recorded the debate one evening and turned it into a whimsical nine-minute minidocumentary, *Puheenvuoro* (Taking the floor). As the drama begins, a frustrated chairwoman, gavel in hand, is begging council members to keep their comments brief. This is the second marathon meeting, she says, and it has been going on for five hours. "I stress that people should not speak just for the fun of it."

Council members, however, continue to demand their turn to speak—their "time on the floor"—to raise questions about the tram. "The train will get stuck in a snowbank and not be able to advance," one woman says, her face etched with concern. "Then the poor people will slosh around in snow."

"Injuries caused by bus tires are always less severe than those caused by tram wheels," another member points out.

An elderly man wearing the white beard of a Russian patriarch uses his time to discuss issues concerning journalism. When the chair interrupts to remind him that the debate is about trams, he responds heatedly, "Yes, the tram or whatever else!" Later, when he is seen wandering up toward the head table, one of his colleagues urges him to return to his seat.

"Even the Siberian flying squirrel thanks us for not building the tram," one woman claims to know.

Eventually, long after the filmmakers turned off their cameras, Tampere's city officials were able to get the council's go-ahead. Work on the tram lines commenced in 2017.

The Tampere council provides an extreme example of democracy's obstacles and delays. But just imagine if this elected body in Tampere, or anywhere else, for that matter, were asked to approve Amazon Prime deliveries by legions of Jeff Bezos's octocopters. They would raise a host of legitimate questions, touching on everything from noise to objects falling from the sky. Matters of air safety would no doubt be far more pressing than the welfare of Siberian flying squirrels.

So where do drones fit into the future of moving our stuff? In most cities, building a business case and winning regulatory approval for fleets of them is likely a ways off. There's little pressing need for shoppers to get their deliveries within minutes, and plenty of elected officials will be all too happy to rail against a corporate behemoth like Amazon and prevent it from taking over the skies.

Yet in different scenarios, an octocopter could be a lifesaver. Imagine one flying over a gridlocked city, but instead of coffee filters or spandex tights, it's delivering a freshly harvested organ to a hospital transplant team. Scientists from the University of Maryland, Baltimore, have been testing just such a system, flying a transplantable kidney three miles in a drone. Legislators and regulators, no doubt, would have fewer quibbles about approving such shipments.

Drones are carrying out similar missions over the densely forested hills of Rwanda and Ghana. A California robotics company, Zipline, operates nests of drones in the two African countries. There they keep fresh supplies of blood. And when they get text messages from rural clinics, whether a woman is hemorrhaging after giving birth or a child is suffering from malaria, health workers load packets of blood onto fixed-wing drones. The drones take off from slingshot launchers and fly to the clinics at about sixty miles an hour—far faster than trucks laboring up and down narrow mountain roads. Upon arrival, Zipline's drones circle briefly above a delivery area, drop their payloads, and fly back. With their round-trip range of ninety miles, according to the company, the drones can serve a rural population of eleven million.

Some decisions are easy for regulators. Drones on medical emergency missions get the green light. A drone delivering a box of Cocoa Puffs? Probably not. But there's an enormous middle ground, scenarios in which delivery drones offer services that are helpful but less than urgent. As these autonomous flying machines proliferate, that's where the debates are sure to heat up. The question isn't whether drones will be on the job, but rather which jobs they'll be allowed to perform.

Conclusion: Minutes, Meters, and Moolah

It's evening rush hour at Rockefeller Center in Midtown Manhattan. Among the crowds walking along Sixth Avenue, stand a few people gazing down at cell phones and then looking anxiously up the street. They've called for a ride-share, but the office towers behind them occupy entire blocks. So it's hard to tell the driver exactly where they are. The trick is to find a landmark small enough for the passenger to be found next to it, but at the same time big enough for the driver to spot. Maybe the taco truck near the corner of Forty-Ninth Street . . .

It has always been a challenge to nail down a specific place on earth and communicate it to someone else. That's because the primitive road maps we use today were fashioned to the specific demands of one preeminent customer: the mail carrier. The primary communication system in each country, the primordial national network, was the one designed to deliver the mail. Even in the days of horse travel, the sprawling army of mailmen had to stop at every street and byway in the entire country, six days a week. They needed roads and streets with names, and each building with a number (and a mailbox). That was the physical order as mandated by the post office. In this scheme, all the vast spaces far from mailboxes, from prairies to beaches, hardly mattered. As far as the mail system was concerned, if a place didn't receive mail, it didn't exist.

When looking for each other in these untagged regions, we often rely on hand and face signals, most of them developed over the millennia. A driver searches the sidewalk crowd for someone with the anxious look of a ride-share passenger. Maybe she'll have an arm up, waving. But would an autonomous car be able to spot one frantically waving passenger across Sixth Avenue from Radio City Music Hall? How do you tell the drone where to pick you up?

Chris Sheldrick was having geography problems. Working in music in London a decade ago, he found it hard to tell delivery companies exactly where to drop off drum sets and speakers. Computers didn't have this problem, he knew. They had a geo-mapping capability and could pinpoint any place on earth by its coordinates. Wilton's Music Hall, the 175-year-old staple in the Whitechapel neighborhood, for example, was at 51.5107° N, 0.0669° W. But try giving a string of numbers like that to an acid rock drummer, and see if he gets to the show.

Sheldrick envisioned a new way to communicate location, whether a favorite oak tree, a buried treasure, a trout-fishing paradise, or a ride-share in Midtown Manhattan. He launched a company to create an address for every spot on the planet.

The challenge was to link the numbered precision of the Global Positioning System with the forgetful brilliance of the human mind. The bridge between the two, he decided, would be language. On a map of the world, he and his partners super-imposed a fine grid. It divided the earth's surface into fifty bil-lion patches, each one three square meters. Then, the tech team set about programming computers to give each of these seg-ments a name that humans could handle.

For this, they did some math. Most of us have somewhere on the order of a 35,000-word vocabulary. If you cube that number, it comes close to fifty trillion. So it was just a matter of getting a computer to combine three words in every conceivable combination—fifty trillion of them. Then they could assign a different name to each room-size parcel on earth, each one in a couple dozen languages. These tags would be easy for humans to remember and exchange with each other, and perhaps most important, to use to communicate locations to our machines. Sheldrick named the company what3words.

The next step was to put the new mapping service on a smartphone app. It would work only if millions of people agreed to try out the standard and download it. Otherwise, it would be as useless as a payment app that only a few stores accept.

As with other ventures we've discussed, from Kevin Czinger's 3-D printing start-up near Los Angeles to century-old giants like Ford Motor Company, there's no guarantee that Chris Sheldrick's mapping company will prevail or even survive. Maybe some other company yet to be founded will gobble up that market. Maybe Google will. But what we can guarantee, whoever comes out on top, is that the world's geography will be crunched, tagged, and organized by computers, and the resulting organization will be communicated to us in a form that we can wrap our fallible minds around. That, likely, will alter the way we think about the places we go.

Throughout this book, we've been discussing the new technologies of mobility and how they're poised to change our cities, our economies, and the fabric of our lives. But they're also likely to change, at fundamental levels, what goes on in our heads—including how we think about space and time.

Consider a day in your life—say, a Saturday. Waking up, you face two key variables. First, time. If it's eight a.m., and you usually go to bed around midnight, you have sixteen hours of waking life, and a body to move one place or another. That raises the second variable: space. Where *can* you go? It's largely a function of the time it takes you to get there, along with the money it will cost. For most of the last century, the answers haven't changed much. But as new mobility options pop up, they scramble the arithmetic of time and space. Places that were beyond reach become accessible. They feel closer.

We've seen this phenomenon before. When the US transcontinental railroad was completed, in 1869, the effect was to dramatically shrink the American continent, from four months to three and a half days. The age of aviation, in the following century, shrank the entire world. How much will the next stage of mobility shrink cities, and how will that change our thinking and behavior?

In the simplest formulation, faster movement saves us time. If an affordable new air service whisks a commuter from Morristown, New Jersey, to New York's Wall Street in fifteen minutes, and spares her the backed-up traffic on I-280 and the teeth-grinding wait at the Holland Tunnel, she might save ten hours a week. It's a gift of time, which she might spend with her family, or taking up a new hobby, or, most likely, working.

But that's assuming that new mobility doesn't change her traveling plans—and that she accepts the new mobility dividend in the currency of time. What if she spends it instead on additional miles?

In the 1970s, an Israeli transportation engineer named Yacov Zahavi studied the patterns of urban travel in cities. He came up with a constant value, the "travel-time measure." It varied from place to place, but the average boiled down to about an hour a day. That was the time human beings would spend going somewhere and coming back every day. In ancient cities we can still visit, from Jerusalem to Venice, people could walk across town, about a mile and a half, in thirty minutes. That was their mobility leash.

Yet when technology advanced to trolleys and later to cars, that leash grew much longer. In theory, people could have harnessed the new mobility to zip to the same old places but in half or a quarter of the time. Instead, though, most tended to expand their horizons and travel farther. They stretched the leash.

An Italian nuclear physicist, Cesare Marchetti, later linked Zahavi's travel-time measure to urban development. In his 1994 paper, "Anthropological Invariants in Travel Behavior," Marchetti described the outlines of cities. He showed how each of these urban footprints expanded as new mobility technologies gained ground. This is hardly news to anyone who has driven the exurban highways of Dallas or Seoul.

So how will Marchetti's collection of urban footprints morph in coming decades? If we can commute, affordably, from a town one hundred miles away in a flying taxi or Hyperloop, do our cities sprawl? Are two cities linked by effective mobility, in effect, the same city? Marchetti posited that a city's borders are defined by our movements. New technologies, he predicted, would fuel the growth of de facto megacities.

"Further growth," we should say, because this process is hardly new. Brooklyn and New York were neighboring cities until the Brooklyn Bridge connected them in 1883. Even though Brooklyn remained a separate city for another fifteen years, it became a functional part of New York the first day it was easy to walk or ride a carriage across the river. In that sense, the new bridge—an engineering advance in mobility—created the megacity of New York.

In the next step, in Marchetti's vision, far bigger conglomerations will occur. New York will extend to Philadelphia and Boston, and beyond. Along the eastern side of China, from Beijing to Hong Kong, he speculated, a megacity of one billion people could emerge.

Only one trouble: this process can also unwind. Fast networks can lose their zip, and as they do, trips that fit neatly into a daily schedule start slowing down. Mayor Eric Garcetti, fondly remembering high school escapades behind the wheel of his 1975 Torino, says, "You could get anywhere in twenty minutes!" Now, he grumbles, the same ride takes an hour.

The phenomenon at the root of this slowdown is known as the Jevons paradox. A central plank of market economies, it states that if something is fabulous, whether a restaurant, a beach, or a highway shortcut, people will flock to it. This growth will continue until the crowding spoils the experience. One sad day you find yourself waiting in a long line to get into a trendy restaurant, or in Eric Garcetti's case, suffering for an hour on what used to be a twenty-minute joyride from West Hollywood to Topanga Canyon. The Jevons paradox stifles growth with increasing degrees of pain and discomfort.

Sadly, the design of our cities, along with our expectations and way of life, are predicated on the old and more efficient status quo. So it's not easy to alleviate the pain of the Jevons paradox. Billions of us are stuck with the ninety-minute commutes, the fifteen-minute crawl to the supermarket.

For the last forty years or so, the twentieth-century auto economy has attempted to defy the Jevons paradox, widening highways, building bridges, and constructing parking lots. This effort is largely fruitless, because each investment invites more motorists—until traffic and parking pinch again, and the entire experience regresses to the vexing mean. It's as if all of us, for decades, have been falling for the same stupid trick.

NOW WE HAVE a once-in-a-century chance for a redo. What we've learned in the Jevons school of hard knocks is that when mobility markets run without constraints, only overcrowding brings the resulting growth to a stop. So how can we foster healthy and fast equilibriums, and keep the wretched Jevons paradox at length? That's where policy comes in.

But here city officials are often confounded by the same phenomenon. They can tweak the infrastructure, for example, adding lanes to inbound traffic during the morning rush hour and reversing them in the evening. But if that speeds up traffic, the Jevons paradox returns, with more traffic piling in until it gets jammed again.

A city can also take steps to discourage driving, by cutting back on parking or narrowing roads to add bike lanes. This may be necessary, but from most drivers' perspectives, it simply shifts the pain from one part of the trip to another. This can leave motorists, still the overwhelming majority of travelers in most US cities, feeling persecuted. (As if their commutes aren't hard enough already!) Their frustrations can nourish political resistance to other mobility initiatives, including public transit.

Drivers make up a constituency that's nearly impossible to please. Perhaps the most effective lever to alleviate hellish traffic jams is financial: to raise the cost of driving. This is the goal of congestion pricing, now well established in cities like London and Stockholm, soon to debut in New York, and possibly even in Los Angeles. When it works, it substitutes money for time, with drivers paying for their saved minutes. London's scheme, which charges nonresidents about fifteen dollars a day to enter the thirteen square miles of the city center, has depressed traffic there by 25 percent over the last decade. Bike usage has spiked.

However, congestion pricing is a blunt-force instrument. Motorists are treated as herds. While the world changes constantly, with traffic ebbing and flowing, a system of rigid rules remains by definition powerless to adapt. Congestion pricing, in that sense, is like the dumb stoplight that for a century has kept us waiting at empty intersections.

Any adjustments in a congestion ordinance are bound to be long and arduous. Say traffic piles up anew in London's Piccadilly Circus. Officials will have to hold meetings, question experts, weigh the testimony of public advocates and bicycling clubs. Finally amid protests, they may vote to raise the daily price by a pound or two. This could take months!

In an ideal scenario, the network would respond to changing conditions in real time, like an organism. Uber provides an early example. With its (highly unpopular) "surge pricing," the rideshare company raises rates, sometimes dramatically, in times of high traffic. It's similar to London's scheme, but much more flexible. The difference is that while London's congestion pricing is fixed on rules, Uber's responds to data. It charges for jams only when they exist.

Such real-time traffic analysis is growing within cities. Already we're seeing the spread of sensors and automatic responses to traffic flows. Smart traffic signals in cities like Stockholm, Dubai, and Hangzhou are an early step in traffic optimization. But there's a long way to go.

For a look at a data-driven future, imagine a variation on Facebook's customer management. The social network processes multitudinous streams of data about its users. It knows what they click on, where they vacation, who their friends are, what music they like. It can approximate their economic status. With this knowledge, the company automatically optimizes its services for its two billion users, serving up to each one the content and advertising he or she is most likely to click on at the most statistically propitious moment. These algorithms are engineered to wring out the most revenue for each customer.

The goal in networked mobility is to build systems, public or private, as responsive as Facebook's. But instead of using streams of data to guide a user along a lucrative string of clicks, such urban systems will attempt to lead each traveler along the most efficient route. Unlike Facebook, which can entrance customers with endless kitty videos or tidbits of celebrity gossip, the traffic models have to move people and cargo in the physical world. It's endlessly more complicated.

It's also focused on a far more complex goal. Facebook's algorithms are optimized to make money. Urban mobility networks, by contrast, must juggle a host of priorities, some of them conflicting. Naturally, they want efficiency—but also safety, fairness, equality of access, air quality, perhaps a smidgen of privacy, and a number of other goals. At the same time, they must allow for fleet operators and bike-share companies to make decent profits, or at least a profit big enough for such outfits to stick around and invest. The job of city regulators will be, increasingly, to monitor these variables. All of them will be encoded into the mobility algorithms running their networks.

Once a mobility data engine is in place, the possibilities are endless. A city might charge drivers minitolls, raising rates on heavy users much the way phone companies treat data hogs. It might grant vouchers to walkers and cyclists and carpoolers. Microtoll rates could fluctuate minute by minute, street by street. In certain cities, customers from rich neighborhoods could pay higher rates, which in turn would subsidize fleets serving mobility deserts. Controversies are bound to erupt. Most of them will center upon algorithms and data.

Many of the burning issues, in fact, will feel like hand-me-downs from the Internet economy. The challenge for cities managing mobility will be to apply the learnings from our misadventures to date in the networked world.

There are three critical issues around data: open standards, algorithm audits, and network neutrality.

Open Standards

The big Internet companies keep oceans of consumer data to themselves. It's their most valuable asset. Cities, however, cannot afford to let mobility companies hoard passenger data in their private stashes. The goal is not to build a dominant player, but instead a vibrant ecosystem.

Mobility information, cloaked for privacy, must be available to everyone, including a company's competitors, so that all parties can build services that feed off each other. A key component for data-sharing ecosystems will likely be distributed-ledger technology such as blockchain. This should enable the various players to share their data broadly without losing ownership of it or disclosing confidential or strategic secrets.

Finland, as we've seen, has come up with a useful model for sharing. By law, every entity providing transport services there, public or private, must make its data available in a common format. This invites service providers, like Sampo Hietenan's Whim, to build mobility services and subscriptions.

Proprietary and cloistered services, whether for data or for charging electronic devices, sap growth from mobility networks.

Algorithm Audits

The Facebook-like algorithms we described earlier in this chapter—the ones that will attempt to micromanage urban circulation—will be fed largely by machine-learning programs. This poses risks. Since such programs feed from historical data, they can perpetuate inequality and unfairness, or even add to it.

Let's say, for example, that a city decides to set up a fleet of autonomous cars to help the elderly and disabled move around. If planners feed historical data from highways and public transit into a computer program, it can detail the patterns of movement and map out the areas where the new service should operate. That sounds useful, and fair.

Just one problem: there might be a poor section of town where many people cannot afford cars and make do with irregular bus service. This is a mobility desert. People there struggle to get to jobs, schools, and even decent supermarkets. They don't move much. So from a data perspective, it might look like a nonentity. If the city's new service is built around this historical data, it will ignore the desert and exacerbate inequality. The people who need help the most will be stuck, and the desert will grow even more barren.

When data-crunching software programs are interpreting the world, these poisonous feedback loops are inevitable. Cathy O'Neil, the author of *Weapons of Math Destruction*, argues that data scientists must lift the lids off the algorithmic black boxes and conduct fairness audits. This will be crucial for mobility networks.

Network Neutrality

Over the past generation, the big cable and phone networks spent heavily to provide bandwidth to billions of customers. Naturally, they're eager to amortize their investments, and they would love to sell premium services to their deep-pocketed customers, including behemoths like Netflix and Amazon. This has led to a battle, in markets around the world, over so-called net neutrality. Should bandwidth be a common good, with equal access mandated for all? Or should Internet providers be free to jam low-rent customers onto stodgy pathways, digital versions of the stifling economy cabins in jetliners?

This same issue is likely to emerge as 5G networks spread to cars. Could a luxury automaker pay a premium for the highest-definition video feed and install more spectacular (and glitch-free) entertainment in their cars and trucks? It could be a selling point. Or should every player get the same service? Regulators and courts will be deciding.

Net neutrality will also play out, bitterly, no doubt, in the physical world, in the markets of time and space. What happens, for example, if an air taxi company invests billions for a network of stations in São Paulo? Can other competitors build businesses on the same stations? City governments will have to decide. Many will likely grant monopolies, at least for a decade or two, so that the first operator can recoup its investment.

Then there's the issue of social equity. What happens if an operator builds stations in São Paulo's rich neighborhoods, offering easy links to the financial strip along Avenida Paulista, but builds no stations in the teeming favelas south and east of the city?

Of course, everyone *should* have access. The theme of equity comes up in every city we visit. Mayors and their transportation staffs often speak about it with passion. But every such demand these cities place on mobility investors amounts to a form of tax. And investors will naturally flock to where the taxes are low ("business friendly" is the term for it). As the mobility revolution spreads, many city governments will find themselves under competitive pressure. It's similar to the jam municipalities face when a beloved sports team threatens to move to a city dangling a $1 billion stadium or a twenty-year tax holiday.

Mobility will bring the entire physical city into play. When companies like Apple or Amazon launch a self-driving fleet in a big chunk of Orlando or Prague, they'll spend heavily, not just on the cars but also on the infrastructure, equipping the streets and traffic signals with all kinds of sensors and cameras. Could Amazon then provide preferential service on this network for its Prime customers? Could Uber lay claim to strategic clusters of curbside real estate?

Such issues will heat up as the Internet economy asserts itself in the physical world. They will pose some of the most contentious political issues for city and national governments. They will define people's neighborhoods, their range of movement, their commutes—in short, their time, space, and economic opportunity.

We don't have clear answers or guiding principles for these questions, other than to suggest that the arrangements should be transparent and limited in time. Mobility, both in technology and governance, is evolving far too fast for any city to commit to long-term deals.

IN THE YEAR 2000, the British government raised $34 billion by selling air. In what was called the biggest auction ever, phone companies bid exorbitantly for slices of the electromagnetic spectrum over the British Isles. Yes, the prices were outrageously rich, but so were the stakes. With so-called 3G networks, cell phones would soon be evolving into Internet machines. The winners would be able to sell a myriad of information services to millions of customers every waking hour. It was a gold rush. Investors were throwing money at anything connected to the so-called Wireless Web.

We were working together as journalists in Paris at the time, covering this upheaval in telecom. It was dramatic, and it raised all kinds of strategic issues. The coming smartphones would blend telephones with consumer electronics and software. So who would lead the industry? Would it be telecom giants from Europe like Nokia, electronics powerhouses from Asia, or American software titans like Microsoft? (Apple, back then, was an afterthought.)

Twenty years later, the parallels with the mobility revolution are striking. Again, we're witnessing a mating process involving massive industries. Software, this time turbocharged with AI, is combining with everything that rolls and flies. Like Wireless Web players at the height of the frenzy, mobility start-ups are swimming in venture funding, many of them vowing to change our lives and our planet, sooner rather than later.

So as we look ahead to the wheeled and winged marvels that await us, it might make sense to study the previous boom, to contrast what was promised with what we ended up with.

Let's start with timing. Consider this lede from our *Business-Week* cover story in the fall of 1999:

> A 15-year-old girl strolls through London's Berkeley Square. Suddenly she hears a beep from her cell phone and looks at the screen. A message sponsored by Starbucks informs her that two friends from her "buddy list" are walking nearby. Would she like to send them an instant message to meet for coffee at the nearest Starbucks around the corner? She merely has to click "yes" on her smart phone to send the message. And she gets an electronic coupon worth $1 off a Frappuccino.

For the most part, the scenario is on point—except that we, along with much of the industry, expected to be experiencing this wireless magic by the following year. Instead we witnessed an industry-wide crash.

Building mobile computers and the networks to feed them, it

turned out, was more complicated than expected. Frustrated early adopters pecked Internet addresses into smartphone browsers and then waited an eternity for soccer scores or stock prices to appear. The promise of video on these small, blurry screens seemed like a cruel joke.

By 2001, it appeared that a functioning mobile Internet market, with all its treasures, might be years away. It wasn't clear which players, if any, would win. Investors exited in a stampede.

What can we learn from that? First, don't trust timetables, whether for flying taxis or hyperloops. Delays are inevitable. This stuff is complicated. Investors are bound to grow weary. The process is sure to chew up many of the players and cast aside others.

That's the bad news. When markets crash, the failures harden into gospel. At the trough of the telecom bust, in 2002, skeptics could recite a litany of the broken promises. Among telecom insiders, our fifteen-year-old girl in Berkeley Square reached emblematic status. She symbolized a giddy industry and, yes, gullible journalists.

Yet five years later, on a January day in 2007, Apple's chief executive, Steve Jobs, stepped to the stage of San Francisco's Moscone Center holding an iPhone. The mobile Internet had arrived. Most people lining up for new iPhones didn't even know that it was five years late. Over the following decade (at the risk of hyperbole), smartphones transformed human communication. Today the mobile Internet is bigger in many ways than the promises of even its most shameless promoters a generation ago.

The takeaway is that the two timelines in a tech boom, investment and technology, rarely move in sync. Every boom starts with hype and feverish hopes. They fuel investment and bring in brainpower. But since they're overblown, they lead more often than not to delays and disappointments. When investors feel burned by bad numbers or technical glitches, they tend to retreat in a thundering herd. The Gartner group goes so far as to chart this process, labeling the low point "the trough of disillusionment."

Yet market panics rarely deliver a telling verdict on a given technology. They're part of the culling process, and they clear the field for the eventual winners.

Compared to smartphones, mobility technology is a big tangle. The phone, for all the engineering brilliance that went into it, was a single technology platform. To this day, the hundreds of millions of units the industry spits out look very much like the original model Jobs held aloft that January day in San Francisco. The transition happened quickly. Within a few years of the iPhone's launch, most of us had traded in our flip phones—and were succumbing to the addiction of pocket-size Internet machines.

The mobility revolution will not steal upon us nearly as quickly. We cannot advance from yesterday to tomorrow simply by purchasing a wondrous artifact. Instead, for most of us, the new order is likely to creep into our lives bit by bit. The skeptics among us—like the fabled frog as the water in the pot slowly grows warmer—will insist that nothing is changing.

Let's conclude this book by sketching out the coming years for a city dweller somewhere in the world. This shouldn't be someone who embraces all the new stuff and brags about it on social networks. No, we'll make her a skeptic, a frog, and watch the scene through her eyes as the water warms.

First, she buys a car. This is the move of a hardened realist, and she has plenty of company. ("All these scooters and whatnot are fine, but I have to actually *get places*.") A year or two later, a new subway line opens. She takes it downtown sometimes. ("Ever since they turned the main lot into a soccer field, it's a pain to park there.") Still, her office is not on the subway line, and she continues to drive to work. ("What mobility revolution?")

She spends a weekend visiting a friend, and she sees that an autonomous fleet is operating in his city. It's limited to a section on the other side of the river. ("Those cars aren't smart enough to go everywhere.") She takes a ride in one and finds it excruciatingly deliberate. ("Boring.")

Friends come back from vacations and talk about hopping across cities in airships. ("Yeah, but not here.") Her cheapskate brother is driving around what looks like the offspring of a motorcycle and a golf cart. ("Good luck when it snows.")

You get the picture. The mobility revolution will continue to insinuate itself into our lives, in some places, and among some people, long before others. Plenty of mishaps and failures along the way will give sustenance to skeptics.

The real test for our hardened realist comes the day she sees, to her horror, a black puddle shimmering under her aging car. It's time to get a new one . . . or is it? Perhaps then she looks around and concludes, despite herself, that she has new choices.

Even at that point, we cannot declare that the new mobility has *arrived*. No, it will never be complete. But epochal changes, in our cities and our lives, will be afoot. In fact, they're already upon us.

ACKNOWLEDGMENTS

When a book like this one takes us to Shanghai and LA and Tampa and Dubai, among many other cities, we have people all over the world to thank. The first person, though, is our wonderful editor, Hollis Heimbouch, who believed in this book from the very beginning, and has provided smart and constant support. Thanks, too, to her colleagues at HarperCollins, including Rebecca Raskin, Nikki Baldauf, and Milan Bozic.

Our agent, Jim Levine, and the team at Levine Greenberg Rostan helped us organize the book and shape the proposal and stayed involved throughout. They also run the friendliest shop on Seventh Avenue (complete with a putting green).

The teams at CoMotion in Los Angeles under Tim Gribaudi, NewCities Foundation in Montreal and Hong Kong, and Olivia Onderdonk provided great support and context.

We also would like to thank a few of the experts who helped us grapple with the technology and the policy challenges ahead. Los Angeles officials, in Mayor Eric Garcetti's office and, especially, Seleta Reynolds's Department of Transportation, provided invaluable help, as did Joshua L. Schank at LA Metro. We're also thankful to Greg Lindsay, Gabe Klein, Boris von Bormann, Grayson Brulte, Arnab Chatterjee, Russ Mitchell, Alan Ohnsman, Bill Visnic, and Marcie Hineman at the Society of Automotive Engineers, and John Dolan and Byron Spice at Carnegie Mellon University.

Others who provided vital help include Lauri Kivinen and Susanna Niinivaara in Helsinki, Noah Raford in Dubai, Lan Shi, Xuetao, and the Shanghai team at Y-CITY Global Innovation Academy, and Miriam Siefer and Joe Bachrach in Detroit. We are grateful for a decades-long friendship with Mohammed Jameel, the chairman of Abdul Latif Jameel in Jeddah, Saudi Arabia—an extraordinary philanthropist and business leader who was among the first to point out that the internal combustion engine would have a finite life.

Finally, this book could not have been researched and written without the support and love of our spouses, Jalaire Craver and Antonella Caruso.

NOTES

INTRODUCTION

xv it will cost about $8,000: American Automobile Association, *Your Driving Costs: How Much Are You Really Paying to Drive?* (Heathrow, FL: AAA Association Communication, 2018), https://publicaffairs resources.aaa.biz/download/11896/.

xvii The need for lithium: Todd C. Frankel, "The Cobalt Pipeline," *Washington Post*, 1 October 2016, https://www.washingtonpost.com /graphics/business/batteries/congo-cobalt-mining-for-lithium-ion -battery/.

xviii But some of them cost only $1,000: Trefor Moss, "China's Giant Market for Really Tiny Cars," *Wall Street Journal*, 21 September 2018, https://www.wsj.com/articles/chinas-giant-market-for-tiny-cars-15375 38585.

xx Little wonder Google: Jon Russell, "Google Confirms Investment in Indonesia's Ride-Hailing Leader Go-Jek," TechCrunch, 28 January 2018, https://techcrunch.com/2018/01/28/google-confirms-go-jek-investment/.

xxi That number is expected: "68% of the World Population Projected to Live in Urban Areas by 2050, Says UN," United Nations Department of Economic and Social Affairs, 16 May 2018, https://www.un.org /development/desa/en/news/population/2018-revision-of-world-urban ization-prospects.html.

xxii By 2030, according to the government's plans: "Dubai's Autonomous Transportation Strategy," Dubai Future Foundation, March 2019, https:// www.dubaifuture.gov.ae/our-initiatives/dubais-autonomous-transportation -strategy/.

xxii And a planned vacuum train: Cleofe Maceda, "Hyperloop in Abu Dhabi to Cost up to Dh1.4 Billion, to Be Ready by 2020," *Gulf News*, 17 January 2019, https://gulfnews.com/business/hyperloop-in-abu-dhabi-to-cost-up -to-dh14-billion-to-be-ready-by-2020-1.1547722844596.

1: HIT ENTER TO PRINT CAR

1 The ports of Los Angeles and Long Beach: "Top 20: U.S. Ports Ranked on 2017 Import Volume," *The Maritime Executive*, 7 June 2018, https://www.maritime-executive.com/article/top-20-u-s-ports-ranked-on-2017-import-volume.

2 Czinger's leading investors: Laurie Chen, "Hong Kong Firm Backed by Li Ka-shing to Build China Plant for 3D-Printed Electric Cars," *South China Morning Post*, 19 June 2018, https://www.scmp.com/business/companies/article/2151288/hong-kong-firm-backed-li-ka-shing-build-china-plant-3d-printed.

3 "For three years": Robert O. Boorstin, "A Tough Pack of Dogs," *Harvard Crimson*, 22 November 1980, https://www.thecrimson.com/article/1980/11/22/a-tough-pack-of-dogs-pif/.

5 Few of these dot-coms: Peter Cohan, "Four Lessons Amazon Learned from Webvan's Flop," *Forbes*, 17 June 2013, https://www.forbes.com/sites/petercohan/2013/06/17/four-lessons-amazon-learned-from-webvans-flop/.

6 He and his team lost control: Alistair Barr, "From the Ashes of Webvan, Amazon Builds a Grocery Business," Reuters, 18 June 2013, https://www.reuters.com/article/net-us-amazon-webvan/from-the-ashes-of-webvan-amazon-builds-a-grocery-business-idUSBRE95H1CC20130618.

7 Within a year: Nichola Groom, "U.S. Electric Car Maker Coda Files for Bankruptcy," Reuters, 1 May 2013, https://www.reuters.com/article/coda-chapter11/u-s-electric-car-maker-coda-files-for-bankruptcy-idUSL2N0DI05J20130501.

7 But in 2009: National Research Council, *Hidden Costs of Energy: Unpriced Consequences of Energy Production and Use* (Washington, DC: National Academies Press, 2010), https://doi.org/10.17226/12794.

8 The first prototype: "Tesla Model S Weight Distribution," *Teslarati* (blog), 19 July 2013, https://www.teslarati.com/tesla-model-s-weight/.

12 Six years later: Chuck Salter, "Barry Diller's Grand Acquisitor," *Fast Company*, 1 December 2007, https://www.fastcompany.com/61073/barry-dillers-grand-acquisitor.

13 In 2017, Arcimoto listed: "Arcimoto Completes $19.5 Million Regulation A+ IPO, Approved for Listing on the Nasdaq Global Market," Business Wire, 21 September 2017, https://www.businesswire.com/news/home/20170921005538/en/Arcimoto-Completes-19.5-Million-Regulation-IPO-Approved.

16 It has the power: Brian Silvestro, "This 700-Horsepower 3D-Printed Supercar Is the Future of Car-Making," *Road & Track*, 27 June 2017, https://www.roadandtrack.com/new-cars/car-technology/a10223824/3d-printed-supercar-future-car-making/.

2: LA: CRAWLING TO TOPANGA CANYON

19 A passenger in Kansas City: Donald Duke and Stan Kistler, *Santa Fe: Steel Rails through California* (San Marino, CA: Golden West Books, 1963).

20 Los Angeles grew: Andre Coleman, "Wealth, Power and Art Are What Drove Railroad Magnate Henry Huntington," *Pasadena Weekly*, 8 June 2017, https://www.pasadenaweekly.com/2017/06/08/wealth-power-art-drove-railroad-magnate-henry-huntington/.

20 In 1911, to meet: "Los Angeles' Auto Manufacturing Past," Los Angeles Almanac, accessed 12 April 2019, http://www.laalmanac.com/transport/tr04.php.

21 As Richard Longstreth describes: Richard W. Longstreth, *City Center to Regional Mall: Architecture, the Automobile, and Retailing in Los Angeles, 1920–1950* (Cambridge, MA: MIT Press, 1997).

21 "So prevalent is": Longstreth, 15.

22 The city has: Neighborhood Data for Social Change, "A 2018 Snapshot of Homelessness in Los Angeles County," KCET, 3 Aug 2018, https://www.kcet.org/shows/city-rising/a-2018-snapshot-of-homelessness-in-los-angeles-county.

22 Parking spaces in greater Los Angeles: Laura Bliss, "Mapping L.A. County's 'Parking Crater,'" CityLab, 11 January 2016, https://www.citylab.com/transportation/2016/01/map-la-county-parking-200-square-miles/423579/.

22 That's five times: Paris covers 104 square kilometers, or 40.7 square miles. See "Capital Facts for Paris, France," World's Capital Cities, accessed 12 April 2019, https://www.worldscapitalcities.com/capital-facts-for-paris-france/.

22 So for a century: Joni Mitchell, "Big Yellow Taxi," track 10 on *Ladies of the Canyon*, Reprise Records, 1970, http://jonimitchell.com/music/song.cfm?id=13.

23 Drivers in LA County: "Miles of Public Roads, Los Angeles County," Los Angeles Almanac, accessed 12 April 2019, http://www.laalmanac.com/transport/tr01.php.

23 The average LA driver: Baruch Feigenbaum and Rebeca Castaneda,

"Los Angeles Has the World's Worst Traffic Congestion—Again," *Los Angeles Daily News*, 19 April 2018, https://www.dailynews.com/2018/04/19/los-angeles-has-the-worlds-worst-traffic-congestion-again/.

25 Now Hawthorne sees: Blanca Barragan, "Take a Tour of the 'Third Los Angeles,' LA's Present and Future," *Curbed Los Angeles* (blog), 14 June 2016, https://la.curbed.com/2016/6/14/11938840/video-third-los-angeles.

27 In 2016, voters: LA Metro details Measure M on its site, http://theplan.metro.net.

28 *Time* magazine named: Lance Morrow, "Feeling Proud Again: Olympic Organizer Peter Ueberroth," *Time*, 5 January 1985, http://content.time.com/time/magazine/article/0,9171,956226,00.html.

29 The region has 6.4 million: "Make Your Car Last 200,000 Miles," *Consumer Reports*, 6 November 2018, https://www.consumerreports.org/car-repair-maintenance/make-your-car-last-200-000-miles/.

29 Only 7 percent: Neighborhood Data for Social Change, "Transit Ridership in Los Angeles County Is on the Decline," KCET, 11 January 2018, https://www.kcet.org/shows/city-rising/transit-ridership-in-los-angeles-county-is-on-the-decline.

29 After decades of progress: Tony Barboza, "Southern California Smog Worsens for Second Straight Year Despite Reduced Emissions," *Los Angeles Times*, 15 November 2017, https://www.latimes.com/local/lanow/la-me-ln-bad-air-days-20171115-story.html.

35 The day after: Noah Smith, "Sudden Appearance of Electric Scooters Irks Santa Monica Officials," *Washington Post*, 10 February 2018, https://www.washingtonpost.com/national/sudden-appearance-of-electric-scooters-irks-santa-monica-officials/2018/02/10/205f6950-0b4f-11e8-95a5-c396801049ef_story.html.

37 In a Vice News feature: Nigel Duara, "People in San Francisco Are Really Pissed over These Electric Scooters," Vice News, 2 May 2018, https://news.vice.com/en_us/article/d35m9a/people-in-san-francisco-are-really-pissed-over-these-electric-scooters.

37 According to Zillow: Chris Glynn and Alexander Casey, "Homelessness Rises Faster Where Rent Exceeds a Third of Income," Zillow Research, 11 December 2018, https://www.zillow.com/research/homelessness-rent-affordability-22247/.

3: 800 ELECTRIC HORSES

44 A study by PwC: "By 2030, the Transport Sector Will Require 138 Million Fewer Cars in Europe and the US," PricewaterhouseCoopers, 16 January

2018, https://press.pwc.com/News-releases/by-2030-the-transport-sector-will-require-138-million-fewer-cars-in-europe-and-the-us/s/a624f0b2-453d-45a0-9615-f4995aaaa6cb.

47 To picture the Darwinian: Trefor Moss, "China Has 487 Electric-Car Makers, and Local Governments Are Clamoring for More," *Wall Street Journal*, 19 July 2018, https://www.wsj.com/articles/china-has-487-electric-car-makers-and-local-governments-are-clamoring-for-more-1531992601.

57 "consumer-friendly charging network": "Our Investment Plan," Electrify America, https://www.electrifyamerica.com/our-plan.

57 But by 2030: "Electric Vehicle Outlook 2018," BloombergNEF, 2018, https://about.bnef.com/electric-vehicle-outlook/.

58 They're mined all too often: "Phone, Electric Cars and Human Rights Abuses—5 Things You Need to Know," Amnesty International, 1 May 2018, https://www.amnesty.org/en/latest/news/2018/05/phones-electric-cars-and-human-rights-abuses-5-things-you-need-to-know/.

62 following the LA Auto Show: Peter Valdes-Dapena, "Amazon Invests Money in Electric Pickups," CNN Business, 15 February 2019, https://www.cnn.com/2019/02/15/business/rivian-amazon/index.html.

4: JURASSIC DETROIT

71 Years later, in 2017: Kirsten Korosec, "Delphi Buys Self-Driving Car Startup NuTonomy for $450 Million," *Fortune*, 24 October 2017, http://fortune.com/2017/10/24/delphi-buys-self-driving-car-startup-nutonomy-for-450-million/.

72 "mobility corridor": Neal E. Boudette, "Ford Aims to Revive a Detroit Train Station and Itself," *New York Times*, 17 June 2018, https://www.nytimes.com/2018/06/17/business/ford-detroit-station.html.

72 Ford, to name one: Kirsten Korosec, "An Inside Look at Ford's $1 Billion Bet on Argo AI," The Verge, 16 August 2017, https://www.theverge.com/2017/8/16/16155254/argo-ai-ford-self-driving-car-autonomous.

73 "We see ourselves": Greg Gardner, "Ford's 'Smart Mobility' Is Still a Long Way from Profitable," *Forbes*, 26 April 2018, https://www.forbes.com/sites/greggardner/2018/04/26/fords-smart-mobility-is-still-a-long-way-from-profitable.

74 Revenue tops $150 billion: "Ford Motor Company Reports Fourth Quarter and Full Year 2017 Results; Revenue Up, Net Income Higher, Adjusted Pre-Tax Profit Lower, Ford Media Center, 24 January 2018, https://media.ford.com/content/fordmedia/fna/us/en/news/2018/01/24/ford-reports-fourth-quarter-and-full-year-2017-results.html.

74 Burly incumbents: Clayton M. Christensen, *The Innovator's Dilemma: When New Technologies Cause Great Firms to Fail* (Boston: Harvard Business School Press, 1997).

79 Asutosh Padhi: "How the Auto Industry Is Preparing for the Car of the Future," *McKinsey Podcast*, December 2017, https://www.mckinsey.com/industries/automotive-and-assembly/our-insights/how-the-auto-industry-is-preparing-for-the-car-of-the-future.

79 Investors like the idea: Kelly J. O'Brien, "Car-Suspension Tech-maker ClearMotion Raises $115M," *Boston Business Journal*, 9 January 2019, https://www.bizjournals.com/boston/news/2019/01/09/car-suspension-tech-maker-clearmotion-raises-115m.html.

83 In the single year: Bernd Heid, Matthias Kässer, Thibaut Müller, and Simon Pautmeier, "Fast Transit: Why Urban E-Buses Lead Electric-Vehicle Growth," McKinsey & Company, October 2018, https://www.mckinsey.com/industries/automotive-and-assembly/our-insights/fast-transit-why-urban-e-buses-lead-electric-vehicle-growth.

84 According to a 2018: Paige St. John, "Stalls, Stops and Breakdowns: Problems Plague Push for Electric Buses," *Los Angeles Times*, 20 May 2018, https://www.latimes.com/local/lanow/la-me-electric-buses-20180520-story.html.

86 A Boston Consulting Group study: "Mobility and Automotive Industry to Create 100,000 Jobs, Exacerbating the Talent Shortage," Boston Consulting Group, press release, 11 January 2019, https://www.bcg.com/d/press/11january2019-mobility-and-automotive-industry-create-jobs-exacerbating-talent-shortage-211519.

5: HELSINKI: WEAVING MAGIC CARPET APPS

90 Over the next several years: Sonja Heikkilä, "Mobility as a Service: A Proposal for Action for the Public Administration, Case Helsinki" (master's thesis, Aalto University, 19 May 2014), https://aaltodoc.aalto.fi/handle/123456789/13133.

97 In 2014, the company: Barb Darrow, "Can We Agree That the Nokia Buy Was a Total Disaster for Microsoft?," *Fortune*, 8 July 2015, http://fortune.com/2015/07/08/was-microsoft-nokia-deal-a-disaster/.

6: IN THE COMPANY OF HAWKS AND NIGHTCRAWLERS

106 Uber has announced: Melissa Repko, "Uber Getting Plans off the Ground for Air Taxis in Dallas, Los Angeles," *Dallas Morning News*,

8 May 2018, https://www.dallasnews.com/business/technology/2018/05/08/uber-getting-plans-ground-air-taxis-dallas-los-angeles.

107 When NASA posted: "NASA Puffin Personal Electric VTOL—Updated Version," YouTube video, posted by "NASAPAV," 16 May 2015, https://www.youtube.com/watch?v=QSdwNl-9mPU.

108 In 2010, the same year: Mark D. Moore, *NASA Puffin Electric Tailsitter VTOL Concept* (Langley, VA: NASA, 2010), https://ntrs.nasa.gov/archive/nasa/casi.ntrs.nasa.gov/20110011311.pdf.

108 After reading Moore's paper: Ashlee Vance and Brad Stone, "Welcome to Larry Page's Secret Flying-Car Factories," *Bloomberg Businessweek*, 9 June 2016, https://www.bloomberg.com/news/articles/2016-06-09/welcome-to-larry-page-s-secret-flying-car-factories.

111 "To get the economics": "Lessons Learned 'Hard'wareway | Uber Elevate," YouTube video, posted by "Uber," 22 May 2018, https://www.youtube.com/watch?v=agnCFyem0kU. This ninety-minute session from the 2018 Uber Elevate Summit, moderated by Mark Moore, delves into the technical and design issues facing eVTOL manufacturers.

115 In 1926, there were fewer: "The Birth of Commercial Aviation," BirthofAviation.org, 12 December 2014, http://www.birthofaviation.org/birth-of-commercial-aviation/.

116 What causes so-called ghost jams: Benjamin Seibold, "Phantom Traffic Jams and Autonomous Vehicles," presentation materials, Temple University, College of Science and Technology Board of Visitors meeting, 21 April 2016, https://cst.temple.edu/sites/cst/files/documents/seibold%20talk%20small.pdf.

118 Musk has come: Bryan Logan, "Elon Musk Is Seeking to Ease Concerns in an Affluent Los Angeles Suburb Where He Wants to Build a Tunnel," *Business Insider*, 18 May 2018, https://www.businessinsider.com.au/the-boring-company-new-test-tunnel-sepulveda-boulevard-los-angeles-2018-5.

7: DUBAI: GRASPING FOR THE CUTTING EDGE

126 Dubai's ruler: Mohammed bin Rashid Al Maktoum, *My Vision* (Dubai: Motivate Publishing, 2012).

129 Instead it would bear: Andy Hoffman, "Dubai's Burj Khalifa: Built out of Opulence; Named for Its Saviour," *Globe and Mail*, 4 January 2010, https://www.theglobeandmail.com/report-on-business/dubais-burj-khalifa-built-out-of-opulence-named-for-its-saviour/article1208413/.

131 The memories of unpaved roads: Jim Krane, *City of Gold: Dubai*

and the Dream of Capitalism (New York: Picador, 2010). Krane's book is an excellent primer on the history of Dubai.

132 it was listed: "Longest Driverless Metro Line," Guinness World Records, 23 May 2011, http://www.guinnessworldrecords.com/world-records/longest-driverless-metro-line.

8: IDIOT SAVANTS AT THE WHEEL

143 In 2015, Uber swooped: Josh Lowensohn, "Uber Gutted Carnegie Mellon's Top Robotics Lab to Build Self-Driving Cars," The Verge, 19 May 2015, https://www.theverge.com/transportation/2015/5/19/8622831/uber-self-driving-cars-carnegie-mellon-poached.

143 A study by Intel: "Intel Predicts Autonomous Driving Will Spur New 'Passenger Economy' Worth $7 Trillion," Intel Corporation, press release, 1 June 2017, https://newsroom.intel.com/news-releases/intel-predicts-autonomous-driving-will-spur-new-passenger-economy-worth-7-trillion/.

149 In a seminal: Charles Picquet, *Rapport sur la marche et les effets du choléra-morbus dans Paris et les communes rurales du département de la Seine* (Paris, 1832), https://gallica.bnf.fr/ark:/12148/bpt6k842918.

155 China alone could account: "Driving into 2025: The Future of Electric Vehicles," J.P.Morgan, 10 October 2018, https://www.jpmorgan.com/global/research/electric-vehicles.

160 This reduces the processing load: Daniel Kahneman, *Thinking, Fast and Slow* (New York: Farrar, Straus and Giroux, 2011).

9: SHANGHAI: PETRI DISH ON THE HUANGPU

166 Acoustic cameras: Alex Hernandez, "Beijing Is Cracking Down on Honking with New Acoustic Camera Systems," Techaeris, 21 April 2018, https://techaeris.com/2018/04/21/beijing-cracking-down-honking-acoustic-camera-systems/.

166 It was a crime: Stephen R. Platt, *Imperial Twilight: The Opium War and the End of China's Last Golden Age* (New York: Alfred A. Knopf, 2018). Platt's history of the Opium War provides rich detail on China's turbulent relations with the West.

169 It took police: Joyce Liu, "In Your Face: China's All-Seeing State," BBC News, 10 December 2017, https://www.bbc.com/news/av/world-asia-china-42248056/in-your-face-china-s-all-seeing-state.

174 At the same time: International Association of Public Transport (UITP), *World Metro Figures* (Brussels: September 2018), https://www

.uitp.org/sites/default/files/cck-focus-papers-files/Statistics%20Brief%20-%20World%20metro%20figures%202018V4_WEB.pdf.

175 Take, for example: Department of City Planning, *Concept Los Angeles: The Concept for the Los Angeles General Plan* (Los Angeles: January 1970), http://libraryarchives.metro.net/DPGTL/losangelescity/1970_concept_los_angeles.pdf.

179 By 2019, Huawei: Nic Fildes, "Equipment Vendors Battle for Early Lead in 5G Contracts," *Financial Times*, 9 October 2018, https://www.ft.com/content/21e34e74-bcf0-11e8-94b2-17176fbf93f5.

180 On the social network Weibo: Echo Huang and Tripti Lahiri, "The Huawei Arrest Fueled Another Online Surge of Outraged Patriotism in China," Quartz, 6 December 2018, https://qz.com/1486133/meng-wanzhous-huawei-arrest-fuels-online-furor-in-china/.

10: SQUADRONS OF PACK DRONES

185 There was no reason: Sara Morrison, "Jeff Bezos' '60 Minutes' Surprise: Amazon Drones," *The Atlantic*, 1 December 2013, https://www.theatlantic.com/technology/archive/2013/12/jeff-bezos-60-minutes-surprise/355626/.

187 A World Bank study: "Jakarta Case Study Overview: Climate Change, Disaster Risk and the Urban Poor: Cities Building Resilience for a Changing World," The World Bank, 2010, https://siteresources.worldbank.org/INTURBANDEVELOPMENT/Resources/336387-1306291319853/CS_Jakarta.pdf.

187 Makarim, as he described: Leighton Cosseboom, "This Guy Turned Go-Jek from a Zombie into Indonesia's Hottest Startup," Tech in Asia, 27 August 2015, https://www.techinasia.com/indonesia-go-jek-nadiem-makarim-profile.

189 Much like Jeff Bezos: Adithya Venkatesan, "How GOJEK Manages 1 Million Drivers with 12 Engineers (Part 2)," Go-Jek blog post on Medium, 1 July 2018, https://blog.gojekengineering.com/how-go-jek-manages-1-million-drivers-with-12-engineers-part-2-35f6a27a0faf.

193 A couple of Finnish filmmakers: *Puheenvuoro* [Taking the floor], directed by Hannes Vartiainen and Pekka Veikkolainen (Helsinki: Pohjankonna Oy, 2017), https://vimeo.com/207997372.

CONCLUSION: MINUTES, METERS, AND MOOLAH

200 He came up with: Thomas F. Golob, Martin J. Beckmann, and Yacov Zahavi, "A Utility-Theory Travel Demand Model Incorporating Travel

Budgets," *Transportation Research, Part B: Methodological* 15, no. 6 (December 1981): 375–89, https://doi.org/10.1016/0191-2615(81)90022-9.

201 In his 1994 paper: Cesare Marchetti, "Anthropological Invariants in Travel Behavior," *Technological Forecasting and Social Change* 47, no. 1 (September 1994): 75–88, https://doi.org/10.1016/0040-1625(94)90041-8.

203 London's scheme, which charges: Nicole Badstuber, "London Congestion Charge: What Worked, What Didn't, What Next," The Conversation, 2 March 2018, http://theconversation.com/london-congestion-charge-what-worked-what-didnt-what-next-92478.

207 Cathy O'Neil, the author of: Cathy O'Neil, *Weapons of Math Destruction* (New York: Crown, 2016).

210 Let's start with timing: Stephen Baker, "Smart Phones: They're the Next Phase in the Tech Revolution, and Soon They May Change Your Life," *BusinessWeek*, 18 October 1999, https://www.bloomberg.com/news/articles/1999-10-17/smart-phones-intl-edition.

INDEX

ABOUT THE AUTHORS

JOHN ROSSANT is the founder and chairman of the NewCities Foundation, a major global nonprofit network devoted to improving the quality of life and work in twenty-first-century cities. He is also the CEO of CoMotion, Inc., an events and media platform focusing on the mobility revolution and the organizer of the well-known CoMotion LA annual conference on the future of mobility. Rossant had previously been responsible for the production of the world's most prestigious international conferences, such as the World Economic Forum Annual Meeting in Davos, Switzerland.

STEPHEN BAKER is the author of five books, including *The Numerati, Final Jeopardy: Man vs. Machine and the Quest to Know Everything*, and a dystopian novel, *The Boost*. For ten years, he was a senior technology writer for *BusinessWeek* in Paris and New York. He previously worked as a journalist in Pittsburgh, Mexico City, Caracas, Madrid, Buenos Aires, and El Paso. He lives in Montclair, New Jersey.